SpringerBriefs in Ecology

SpringerBriefs present concise summaries of cutting-edge research and practical applications across a wide spectrum of fields. Featuring compact volumes of 50 to 125 pages, the series covers a range of content from professional to academic.

Typical topics might include:

- A timely report of state-of-the art analytical techniques
- A bridge between new research results, as published in journal articles, and a contextual literature review
- A snapshot of a hot or emerging topic
- An in-depth case study or clinical example
- A presentation of core concepts that students must understand in order to make independent contributions

More information about this series at http://www.springer.com/series/10157

Derek France · W. Brian Whalley
Alice Mauchline · Victoria Powell
Katharine Welsh · Alex Lerczak
Julian Park · Robert Bednarz

Enhancing Fieldwork Learning Using Mobile Technologies

Derek France
Department of Geography
and International Development
University of Chester
Chester
UK

W. Brian Whalley
Department of Geography
University of Sheffield
Sheffield
UK

Alice Mauchline
School of Agriculture, Policy
and Development
University of Reading
Reading
UK

Victoria Powell
Department of Geography
and International Development
University of Chester
Chester
UK

Katharine Welsh
Department of Geography
and International Development
University of Chester
Chester
UK

Alex Lerczak
Department of Geography
and International Development
University of Chester
Chester
UK

Julian Park
School of Agriculture, Policy
and Development
University of Reading
Reading
UK

Robert Bednarz
Department of Geography
Texas A&M University
College Station, TX
USA

ISSN 2192-4759 ISSN 2192-4767 (electronic)
SpringerBriefs in Ecology
ISBN 978-3-319-20966-1 ISBN 978-3-319-20967-8 (eBook)
DOI 10.1007/978-3-319-20967-8

Library of Congress Control Number: 2015943798

Springer Cham Heidelberg New York Dordrecht London
© The Author(s) 2015
This work is subject to copyright. All rights are reserved by the Publisher, whether the whole or part of the material is concerned, specifically the rights of translation, reprinting, reuse of illustrations, recitation, broadcasting, reproduction on microfilms or in any other physical way, and transmission or information storage and retrieval, electronic adaptation, computer software, or by similar or dissimilar methodology now known or hereafter developed.
The use of general descriptive names, registered names, trademarks, service marks, etc. in this publication does not imply, even in the absence of a specific statement, that such names are exempt from the relevant protective laws and regulations and therefore free for general use.
The publisher, the authors and the editors are safe to assume that the advice and information in this book are believed to be true and accurate at the date of publication. Neither the publisher nor the authors or the editors give a warranty, express or implied, with respect to the material contained herein or for any errors or omissions that may have been made.

Printed on acid-free paper

Springer International Publishing AG Switzerland is part of Springer Science+Business Media (www.springer.com)

Preface

Recent advancements in hardware, software, battery life and durability of portable devices have led to the development of tablet computers and other associated mobile technologies. Tablet technology only became available to the mass market since 2010, and this has forced us to think afresh about how we educate and get the best from our students. Indeed, how we can support students to use affordances, in the sense that a pencil and paper are affordances, to get the best from themselves. We believe that 'tablets' such as Apple iPads, Samsung Galaxies, Xperia tablets, etc., can be viewed as transforming devices at all levels within education and especially within fieldwork education. We would even say that they are a 'disruptive innovation' within education as they create a new market and eventually displace existing technology, thereby creating new educational possibilities.

Fieldwork is a core element of many Bioscience, Geography, Geology, Earth and Environmental Science degree courses. Fieldwork can provide opportunities for experiential learning and research-led teaching in a 'real-world' setting. Teaching and learning during fieldwork can be enhanced through the use of digital technologies; tablets provide opportunities to develop novel approaches to fieldwork pedagogy that neither students nor tutors envisaged possible through traditional means.

The aim of this book is to help you as a tutor to develop novel pedagogic approaches that make the most of these new digital technologies to enhance fieldwork teaching and learning. There is a supporting website that continues to be updated as new technologies and pedagogic ideas emerge: http://www.enhancingfieldwork.org.uk.

Acknowledgments

The authors would like to acknowledge and thank the Higher Education Academy of the UK for its support and funding of the 3-year National Teaching Fellowship Project entitled 'Enhancing Fieldwork Learning'. Continuation funding from The British Ecological Society has allowed the team to continue to develop and share good practice in fieldwork teaching.

The above project has been a catalyst for practitioners to contribute and share good practice case studies of Technology-Enhanced Learning in the field. We would like to thank all those practitioners for sharing their practice, the students who participated in the research projects, all those students and colleagues who responded to surveys and workshop participants for their enthusiasm and input into the project.

The authors and publisher would also like to thank the following practitioners for their good practice case studies:

Andrew Goodliffe, Ian Stimpson, Liz Earley, Maryanne Wills, Gary Priestnall, Claire Jarvis, Jennifer Dickie, Gavin Brown, Rob Jackson, Charles Harrison, Paul Wright, Rhu Nash, David Aanensen, Derek Huntley, Edward Feil, Fada'a al-Own, Brian Spratt, Elizabeth Fitzgerald, Meg Stewart, Jeffrey Clark, Jeremy Donald, Keri Van Camp, Ian Fuller, Susan P. Mains, Lorraine van Blerk, Jade Catterson, Ming Nie, Claire Jarvis, Jennifer Dickie, Sandy Winterbottom, Peter Bunting and Carina Fearnley, Rebecca Thomas, Servel Miller, Judith E. Lock, Moira A. Maclean, Chris D. Sturdy, Davide Zilli, Alex C. Rogers, Liam Basford, Stephen Birch, Alison Black, Alastair Culham, Hazel McGoff, Karsten Lundqvist, Philippa Oppenheimer, Jon Tanner, Mark Wells, Liz White, Phil Porter, Varyl R. Thorndycraft, Don Thompson, Emily Tomlinson, Tim Stott, Anne-Marie Nuttall, Jim McCloskey, Richard J. Stumpf II, John Douglass, Ronald I. Dorn, Trevor Collins, Jim Wright, Sarah Davies, Jessica Bartlett.

These practitioners contributed 29 case studies to this book and we describe over 130 mobile apps with suggestions on their usage in enhancing fieldwork learning. This book is not intended to be an exhaustive list of mobile apps applicable to fieldwork, but used to illustrate the diverse range available and their educational potential within a fieldwork setting.

Contents

1	**Introduction**	1
1.1	Fieldwork and Experiential Learning	1
1.2	Integration of Technology and Mobile Devices into Everyday Life	2
1.3	Possibilities of Using Mobile Technologies in Education	3
1.4	Mobile Learning	4
1.5	Adoption of Mobile Device Use in Higher Education	4
1.6	Limitations and Concerns with the Use of Mobile Devices	5
1.7	Personal Learning Environments and Personal Learning Networks	5
1.8	Potential for Enhancing Fieldwork Activities Using Mobile Devices	8
1.9	How This Book Is Organised	10
1.10	Summary	13
	References	14
2	**Introduction to Tablets and Their Capabilities**	17
2.1	Usability of Tablets	17
2.2	Getting Used to a Device	20
2.3	Communication	21
2.4	Accessibility	21
2.5	Connectivity	22
2.6	Digital Mapping	22
2.7	Data Storage	23
2.8	Data Sharing	23
2.9	Data Processing	24
2.10	Data Collection and Loggers	25
2.11	Summary	26
	References	26

3	Capturing and Using Visual Imagery in the Field	27
	3.1 Still Visual Imagery on Tablets	27
	3.1.1 Examples of Use—The Basics	27
	3.2 Drawing and Painting Tools	33
	3.3 Summary	46
	References	46
4	Display and Recording: e-Books and Field Notebooks	47
	4.1 Fieldwork Materials	47
	4.2 e-Books and Related Formats	48
	4.3 PDF Manipulation	48
	4.4 Field Notebooks	49
	4.4.1 Livescribe Smartpen	49
	4.5 Geo-referencing and Geo-tagging	52
	4.5.1 Problems and Limitations with Geo-referencing	53
	4.6 Summary	64
	References	64
5	Utilising Video in Fieldwork	65
	5.1 Video Recording	65
	5.2 Podcasting	75
	5.3 Top Tips for Better Videos	83
	5.4 Summary	83
	References	84
6	Social Networking, Communication and Student Partnerships	85
	6.1 Introduction	85
	6.2 Social Networking	86
	6.3 Communication	92
	6.3.1 Personal/Audience Response Devices	92
	6.3.2 QR Codes	93
	6.3.3 Word Clouds	93
	6.4 Student Participation	94
	6.5 Summary	98
	References	99
7	Pre-field Trips and Virtual Field Trips	101
	7.1 Introduction	101
	7.2 Re-usable Learning Objects and Fieldwork	102
	7.3 Augmented Reality	106
	7.4 Summary	113
	References	114

8	**Portable Networks and Specialised Fieldwork Applications**		**115**
	8.1	Introduction	115
	8.2	Portable Networks: Remote Access Through Local Wi-Fi Services	115
	8.3	Photography	127
		8.3.1 Microscopes	127
		8.3.2 Telescopes	128
		8.3.3 GigaPan	129
		8.3.4 Hero Camera	130
	8.4	Surveying Equipment	131
	8.5	Specialised Fieldwork Apps	133
		8.5.1 Distance and Angle Measurement	133
		8.5.2 Geological Angle Measurement	134
		8.5.3 Light Measurement	134
		8.5.4 Temperature Measurement	135
		8.5.5 Magnetic Field Measurement	135
		8.5.6 Sound Measurement	135
		8.5.7 Time Measurement	135
		8.5.8 Accelerometer	135
		8.5.9 Calculators	135
		8.5.10 Geoscientists	136
		8.5.11 Species Identification Guides and Citizen Science Apps	136
	8.6	Summary	137
	References		137
9	**Conclusions and Recommendations**		**139**
	9.1	Conclusions	139
	9.2	Technology and Employability	142
	9.3	Bring Your Own Device (BYOD)	142
	9.4	The Enhancing Fieldwork Learning Project—Conclusions	143
	9.5	Top Tips for Tutors	145
	9.6	Final Comments	146
	Further Resources		147
	References		148
List of Mobile Apps Mentioned			**151**
Index			**153**

Abbreviations

3G/4G	Third or fourth generation mobile network
App	Mobile application—software designed to run on mobile devices and smartphone
AR	Augmented reality
BT	Bluetooth—wireless technology for exchanging data over short distances
BYOD/T	Bring Your Own Device/Technology
CSV	Comma separated values
EFL	The Enhancing Fieldwork Learning project
ERA	Enabling Remote Activity
FNS	Field Network System
FSC	Field Studies Council
GA	Graduate Attribute
GEES	Geography, Earth and Environmental Sciences
GLE	Group Learning Environment
GPS	Global Positioning System
HE	Higher Education
HEA	Higher Education Academy
ICT	Information and Communications Technology
KML	Keyhole Markup Language
LiDAR	Light Detection and Ranging
MDMD	Mobile Data Mini Directories
MO	Media Object
OER	Open Educational Resource
OU	Open University
PDA	Personal Digital Assistant Computer
PDF	Portable Document Format—a file format that can be used to save documents that look the same in any operating system
PLE	Personal Learning Environment
PSP	PlayStation Portable

RLO	Re-usable learning object
SETT	SETT Framework—Student, Environment, Task, Tools
TEL	Technology Enhanced Learning
USB	Universal Serial Bus—used for connection between computers and other devices
VFG	Virtual Field Guide
VGA	Video Graphics Array—PC connector for a projector
VR	Virtual Reality
VTF	Virtual Field Trip
XML	Extensible Markup Language

List of Figures

Figure 1.1	Interaction of a personal learning environment (PLE) within a variety of educational spaces (taken from France et al. 2013)................................	6
Figure 1.2	Basic student-engagement relationships after Beetham (2013)	7
Figure 1.3	The relationships from Fig. 1.2 associated with personal learning environments from Fig. 1.1.................	7
Figure 1.4	The FRAME, model of Koole (2010) showing the convergence of technologies, learning capabilities and social interaction	9
Figure 2.1	A word cloud bringing together the main comments from a questionnaire after 'easy' and 'useful' had been deleted	18
Figure 2.2	iPad Mini in Griffin 'Survivor' case.................	18
Figure 2.3	Word cloud formed from students responding to 'disadvantages' in a post field trip questionnaire on the use of iPads in the field..................................	19
Figure 2.4	iPad Mini with BT keyboard acting as a rest. Below is the holding part of a case with the screen flap turned back and key ring holding an extending safety line. To the *left* is an Apple BT full size keyboard, BT earphones and microphone and the silver 'drum' is a desk-top BT loudspeaker...	19
Figure 3.1	*Left*, field sketch using *Skitch* on its own, *right photograph* with annotation by *Skitch* done in the field and used as a basis for the sketch	34
Figure 3.2	Using Skitch to reflect on recreational spaces (Photograph courtesy of Liz Earley)............................	36
Figure 3.3	Part of a *Skitch* to reflect on stream characteristics pinned to a Google Earth screen shot (Image courtesy of Liz Earley. Map data: Google, Get Mapping plc)	37

Figure 4.1	Screengrab of sample locations recorded using *GPS log* then plotted and shared using *Geospike*. (Map data: Google, TerraMetrics)	51
Figure 4.2	Students learning how to use GIS technology in the field. (Image courtesy of Stewart, Clark, Donald and Van-Camp)	61
Figure 4.3	An example of a soil lead concentration interpolated map around Appleton. (Image courtesy of Stewart, Clark, Donald and VanCamp)	62
Figure 5.1	Students surveying the glacial foreland at Fox glacier, New Zealand	70
Figure 6.1	Word Cloud from qualitative data from missionary diaries in Madagascar 1865–1900. (Courtesy of Prof. David Nash, University of Brighton)	94
Figure 8.1	The field network system (FNS), comprising a small rucksack, battery powered Wi-Fi router and laptop (*left*), which are used to create a local network and website to support inquiry-based fieldwork learning activities using a web-standards compliant browser on mobile devices (*right*). Photographs courtesy of Trevor Collins	117
Figure 8.2	The ERA toolkit in use on the OU's environmental change course at Howick Haven, Northumberland in 2009. The students (*left*) are discussing the site's geology with the tutor (*right*) while watching a video feed from the tutor's helmet camera. Photos taken by tutor are displayed on the students' second screen (A video clip is available at YouTube 2014). Photographs courtesy of Chris Valentine and Trevor Collins (both were first published in Collins et al. 2010)	122
Figure 8.3	Photograph of a lichen on a rock taken in the field using a *ProScope* attached to an iPad Mini	128
Figure 8.4	DSLR camera with long-focus zoom lens on a *GigaPan* frame	129
Figure 8.5	Annotated *Google Earth* vertical image with the origin of a *GigaPan* view. This could be built into a website to provide more information or perhaps a set of associated questions. (Map data: Google, Tele Atlas)	130
Figure 8.6	The Hero camera (*right*) with SDHC card, *centre*, waterproof case and *left*, bracket and camera tripod fitting. Photo using the on-board camera of an iPad. The card's images (.mov) can be read from the card through a card reader onto the iPad and processed as required from an app such as *Splice*	131

Figure 8.7	Screen shot of an iPad photograph with the overlaid information from the surveying app *Theodolite*	132
Figure 8.8	A *Google Earth* image with DGPS-surveyed profiles of a river (under the tree canopy). This image was shared between the groups doing the surveying. (Map data: Google)	132
Figure 9.1	Ways that mobile devices can facilitate the interaction between PLEs and GLEs via cloud storage and Web 2.0 interactivity	140
Figure 9.2	Undergraduate skills-to-practices development schema modified after Sharpe and Beetham (2010) by adding a top tier, Team development' and with mutiple facets-skills	141

Chapter 1
Introduction

Abstract This Chapter introduces the concept of fieldwork and the literature promoting the value of integrating mobile technologies into fieldwork practice. We discuss the pedagogic reasons for incorporating technology into fieldwork and promote the concept of Personal Learning Environments (PLEs) for students as a framework for education, with the "tablet" computer (loaded with appropriate apps) enhancing the student's personal learning experience.

Keywords Fieldwork · Mobile device · Personal learning environment · Pedagogy · Technology-enhanced learning

1.1 Fieldwork and Experiential Learning

Fieldwork can often be thought of as working in extreme environments for long periods, we prefer the broad definition of any 'out of classroom/lecture theatre educational experience'. Fieldwork can enhance students' experience of whatever they are studying (France and Ribchester 2004; Hovorka and Wolf 2009; Scott et al. 2012; Welsh et al. 2013; Wheeler et al. 2011) and provide ample opportunities for experiential learning or learning from experience (Keeton and Tate 1978; Kolb 1984). We believe that experiential learning is at the heart of the best fieldwork practices (Healey and Jenkins 2000).

Fieldwork can offer students a novel learning environment (Rickinson et al. 2004; Cotton and Cotton 2009), a valuable learning experience (Fuller et al. 2006) and provide opportunities for experiential learning which can motivate (Parr and Trexler 2011) and benefit students that find other teaching methods less rewarding (Kern and Carpenter 1984, 1986). Recent studies have provided evidence of enhancement of learning during fieldwork (Prokop et al. 2007; Easton and Gilburn 2012; Scott et al. 2012) and much of this enhancement has been attributed to the fact that most students enjoy and value fieldwork (Gamarra et al. 2010; Goulder et al. 2012).

Students have the opportunity to develop discipline-specific practical skills during fieldwork in addition to enhancing their wider, employability skills such as team working, the development of interpersonal skills, self-management and lifelong learning skills (Andrews et al. 2003). There is even the prospect to provide a new stage for skills development (France and Ribchester 2004). Personal development is especially evident when away on residential fieldtrips (Stokes and Boyle 2009).

Fieldwork is undertaken in a wide range of subject disciplines and with a pedagogic remit, but in general we want to show students how to observe the natural world, to interpret it, to collect data and samples, to analyse data and compare trends and results. It is all about giving the students a chance to immerse themselves in the environment and to make their learning active (Swansborough et al. 2007).

In order to improve and enhance fieldwork learning there is always the potential to innovate. The underlying rationale for introducing any educational device or technique needs to be pedagogy. This can be exemplified if you are innovating in order to find ways to:

- Deliver a better learning experience for your students
- Do something you can't do now, or want to do it better
- Improve the group learning dynamic on fieldwork

In this book we present a wide variety of ways to use tablets and mobile devices. However, it is for you, the tutor, to think of how best to use them in your own educational situations. Section 1.3 is given to a basic pedagogic consideration of using technology for effective educational practice. However, fieldwork design and planning is just as important as where it is carried out and how long it takes. Further guidance on planning fieldwork can be found for the Biosciences in Peacock et al. (2011) and for Geography Earth and Environmental Sciences (GEES) in Maskall and Stokes (2008).

1.2 Integration of Technology and Mobile Devices into Everyday Life

Technology is now integrated into everyday life and study. Weiser (1991) in his article on the 21st Century Computer described this as 'ubiquitous computing' where "the most profound technologies are those that disappear—they weave themselves into the fabric of everyday life until they are indistinguishable from it" (Weiser 1991).

As we are now in the age of pervasive or ubiquitous computing (Caudill 2010; Traxler and Wishart 2011; Mehigan and Pitt 2010) the public's expectancy is for researchers and educators to keep up with the pace of mobile technology advancement (Traxler and Wishart 2011). Indeed, students have access to handheld or mobile technologies that are more powerful and better connected than most conventional desktop computers (Guy 2010).

'Mobile device' is a term now commonly used to mean any device that can record, transfer or provide information to the user, in any location (Masrom and Ismail 2010). The functions of these devices are most important; allowing the transfer of information over wireless and Bluetooth connections, capturing and storing multimedia data along with a variety of analytical applications (Traxler and Wishart 2011; Shih and Mills 2007).

Smartphones, tablets and other mobile devices are designed around the market's needs, following trends and meeting a business case for the provision of specific features (Caudill 2010). This has led to diversity, fast progression of technology and out-modding of Smartphone models (Traxler and Wishart 2011). This trend is particularly illustrated by Caudill (2010):

> Moore's Law, which states that processing power at a given price will double approximately every two years, has already been eclipsed by the rapid pace of hardware advancement. Technology devices are steadily becoming more and more capable and less and less expensive over time. This trend can be expected to continue, and perhaps quicken, as each new advance builds on the already existing technology.

1.3 Possibilities of Using Mobile Technologies in Education

Such rapid advancement in the development of mobile devices has led to huge diversity; everything, from the size of the keyboard to the availability of apps, varies from device to device. This diversity of devices has led to confusion and incoherence with regards to building a standard structure for pedagogical device use (Traxler and Wishart 2011). Yet the potential pedagogic benefits are immense. These advances and integration into pedagogy are perceived to be an even greater asset to flexible learning that the initiation of distance learning (Masrom and Ismail 2010).

Mobile devices offer incredible potential for mobile learning, with particular benefits from:

- Touch screens
- Supportive and flexible user interfaces
- Disability-friendly 'gestures' and interfaces
- Integrated GPS
- Connectivity via Wi-Fi and 3G/4G
- Artificial intelligence
- Augmented reality
- Integrated cameras and audio recording
- Non-restrictive operating systems, such as Android, for users to design their own apps

1.4 Mobile Learning

Universities now have the opportunity to move away from a model of fixed, dedicated general computing spaces towards a mobile, wireless computing paradigm that turns any space into a potential learning space. Mobile learning is a term used to describe the use of e-learning on the move (Jarvis and Dickie 2010). The concept of mobile learning constantly develops, adapting to new technological advances and pedagogical possibilities.

Given the possibilities mobile learning enables, it actually means *more* than just 'e-learning on the move'; it opens up the possibilities for students to interact in a three-way connection between themselves, learning resources and a global-social network (Traxler and Wishart 2011). It offers "here and now learning", leading the learning environment into the field and laboratory (Martin and Ertzberger 2013).

The benefits to students through mobile learning at higher education include:

Cochrane and Bateman (2010)

- Exploring innovative teaching and learning practices
- Enabling the embodiment of 'authentic learning'—i.e. facilitating anywhere, anytime, student centred learning
- Engaging students with the affordances of mobile Web 2.0 technologies: connectivity, mobility, geolocation, social networking, personal podcasting and vodcasting, etc.
- Bridging the 'digital divide' by providing access to learning contexts and user content creation tools that are affordable and increasingly owned by students

Masrom and Ismail (2010)

- Availability of performance support through interactive materials and guidance
- Communication access to experts
- Ability to learn whenever and wherever

1.5 Adoption of Mobile Device Use in Higher Education

Smartphone use in secondary and tertiary education has previously been perceived by teachers to have anti-authoritative and destructive connotations. This might lead, potentially, to distractions, game-playing and cheating (Campbell 2006; Wei and Leung 1999; Katz 2005).

However, these devices and tools are now seen as having enormous potential for information transportation and engagement (Knight 2006). Staff can have the most significant impact on the adoption of mobile technologies for learning, offering the potential for enhancement or hindrance to integration methods (Mac Callum and Jeffrey 2010). In fact, Knight (2006) stressed that when developing technology-supported teaching methods, staff must ensure that they first fully understand their students':

- technical ability to make best use of the technology available to them
- attitude to the use of technology, including their prior knowledge and experiences
- access to personal technologies, including their own PC, mp3 player and mobile phone

Mobile devices can open up a range of different scenarios to students, opening up the accessibility to learning (Masrom and Ismail 2010).

1.6 Limitations and Concerns with the Use of Mobile Devices

Masrom and Ismail (2010) also identified barriers to learning through mobile devices; including technological constraints, fragmentation of the learning experience, limited development of the learners ability to control their own learning experience, screen size, cost and security concerns. Students also worry about the potential for damage to the institution's or their own equipment (Beddall-Hill et al. 2011).

Another concern lies with the potential for privacy to be compromised as students use their own devices and private accounts (Knight 2006). Covert surveillance of students while using social network tools for research and assignment participation needs to be investigated further to ensure student safety (Beddall-Hill et al. 2011). Providing supervised Institutional devices and accounts can alleviate some of these concerns, but anonymisation in an unsupervised environment can produce additional security problems.

1.7 Personal Learning Environments and Personal Learning Networks

We now take some of the points mentioned in the previous sections, mainly concerned with mobile devices, and consider them in the general context of enhancing fieldwork learning. We have been promoting the concept of Personal Learning Environments (PLEs) for students as a framework for education (Fig. 1.1),

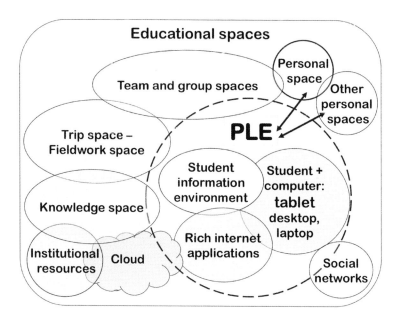

Fig. 1.1 Interaction of a personal learning environment (PLE) within a variety of educational spaces (taken from France et al. 2013)

especially with Technology-Enhanced Learning (TEL). We define Personal Learning Environments as a location where a learner happens to be (Whalley et al. 2014). Learners will move around within educational spaces as shown in Fig. 1.1. This concept of PLEs is supported by tablet technology such that a tablet can, with appropriate apps, enhance the learning experience by allowing students to take full advantage of truly personalised computing.

We also want to encourage student involvement or 'engagement' in their learning and the following schema (Fig. 1.2 after Beetham 2013) provides such a rationale. The tasks shown in Fig. 1.2 are seen in many fieldwork and out-of-classroom activities. The learning activities do not necessarily require network environments as communication can be asynchronous, depending on local conditions and connectivity. The 'reporting' aspect (communicating and writing notebooks) has also been added to Beetham's schema in Fig. 1.2 as we view that, especially for fieldwork, this is an important activity for students.

In Fig. 1.3 we bring the concept of Personal Learning Environments together with the aspects of student engagement together in a general scheme for exploiting technology with the focus on student experience and performance. We suggest that these frameworks provide ways of viewing new fieldwork practice and reviewing existing practice. This also demonstrates that the benefits provide by mobile computing and technology can be placed centrally within an educational or pedagogical framework; moreover this scheme focuses on learning activities.

1.7 Personal Learning Environments and Personal Learning Networks

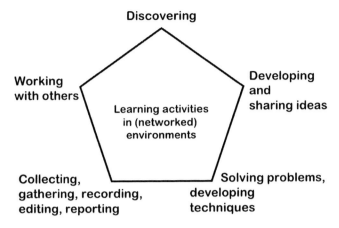

Fig. 1.2 Basic student-engagement relationships after Beetham (2013)

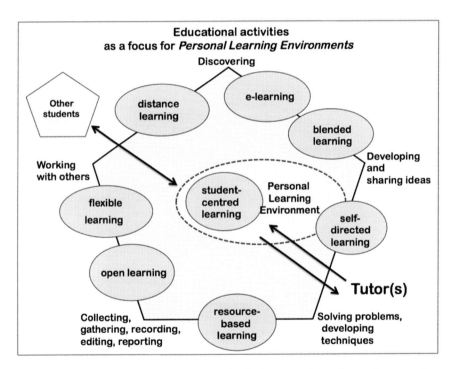

Fig. 1.3 The relationships from Fig. 1.2 associated with personal learning environments from Fig. 1.1

1.8 Potential for Enhancing Fieldwork Activities Using Mobile Devices

Although smartphones are becoming more affordable and ubiquitous (Melhuish and Falloon 2010) a study of undergraduate students found that many who own smartphones were largely unaware of their potential for use in their own education (Woodcock et al. 2012). It is therefore the responsibility of the educators to help students to realise this potential by designing educational materials and activities with this in mind.

Similarly, tablet computers are a key emerging technology in Higher Education (Johnson et al. 2013) and the first International Conference on the use of iPads in Higher Education was held in 2014 (ipadsinHE 2014) as a critical forum for sharing examples of instructional applications in HE as well as discussing innovative pedagogies. This kind of discussion forum allows the development and sharing of novel pedagogic approaches across discipline boundaries.

Fieldwork teaching and learning can certainly be enhanced through the use of technology. Welsh et al. (2013) conducted an international survey of HE fieldwork practitioners and found there was high technology usage before and after time in the field, but some early adopters were also using portable devices such as smartphones and global positioning systems while out in the field. The main pedagogic reasons for cited for the use of technology were the need for efficient data processing and for skill development (see Table 1.1).

The conclusions from the survey were that "fieldwork practitioners are passionate about the subject they teach, but the survey results showed that the main, generic, pedagogic problem encountered during fieldwork teaching is data processing (the speed and ease of data collection and analysis), which can be easily solved through the use of ubiquitously available technology."

This book has focused on the tablet computer as a good device to support fieldwork teaching and learning. The previous generation of small transportable devices, PDAs Personal Digital Assistants (palmtops) have evolved into tablets. We believe that 'tablets' such as Apple iPads, Samsung Galaxies, Xperia tablets etc.

Table 1.1 Main pedagogic reasons for introducing technology to fieldwork (Welsh et al. 2013)

	Tutors' most important reasons for introducing technology to fieldwork teaching (from 2011 survey)
1	Data processing; faster, easier, storage, security, sharing and analysis in field, improved accuracy
2	Skill development; general ICT skills, subject-specific technologies e.g. GPS, employability
3	Post-fieldwork revision/reflection/reporting
4	Enhancing the learning experience
5	Facilitate communication; between students in the field, between countries, between field and 'base', for safety

1.8 Potential for Enhancing Fieldwork Activities ...

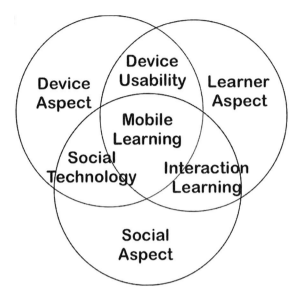

Fig. 1.4 The FRAME, model of Koole (2010) showing the convergence of technologies, learning capabilities and social interaction

can be viewed as transforming devices at all levels within education and especially within fieldwork education. We would even say that they are a 'disruptive innovation' within education (Christensen et al. 2008) as they create a new market and eventually displace existing technology thereby creating new educational possibilities. We refer here mainly to Apple products; this is mainly because we have used them extensively on our Enhancing Fieldwork Learning project from the outset. Alternative operating systems to Apple's iOS are available, most notably, Android. However, the operating system has little effect on the basic device and how it is used.

Therefore this book takes this opportunity to enhance experiential field education by sharing examples of ways to integrate technology. We would like to help you to think about your fieldwork or out-of-class activities and make them as experiential as possible. Tablet technology can help you do this and through this book we want to share our experiences (and those who have provided case studies) with you and your students.

We leave the pedagogy to you, for you to decide what you want to do and we hope the case studies in this book will give you some ideas. We provide some ideas and hints that are pedagogically sound. We would also urge you to browse JISC's 'Design Studio' (JISC 2014) for ideas and examples of successful implementations.

Within the basic pedagogic frameworks summarised in Fig. 1.3 we have also used the Framework for the Rational Analysis of Mobile Education (FRAME) model (Koole 2009) (Fig. 1.4) to guide the development of use of apps and devices in the project. Although this model was developed before the release of the iPad it still holds good and in fact is reinforced by the development of apps, especially those involved with social technology and interaction learning.

1.9 How This Book Is Organised

The structure of this book is associated with our senses; about how we experience and learn from the world around us. We want to show that tablets can be used in a wide variety of ways and that the physical aspects of tablets utilise apps that often link these senses together. So we have categorised according to 'sense' (Table 1.2); with brief discussions about tablets' capabilities, suggestions of apps that use these senses followed by some examples and case studies. Some case studies use more than one app and educational method or intervention but follow, as far as possible, within the main headings. Hence it is possible to dip into and browse the book in order to search for ideas rather than read it through cover-to-cover.

Table 1.2 is a general scheme showing the basic facilities provided by most mobile devices. We cover most of these senses/modes in the following Chapters and individual case studies may cover more than one aspect of functionality.

We concentrate on tablets, rather than Smartphones, in this book as they provide a basic capability of portability, screen size, memory capacity and battery life; what might be called 'usability'. This does not mean that smartphones cannot be used but their small screen size may be less than useful; these aspects are discussed later. Wi-Fi connectivity can be useful but may not be essential although some apps will rely on Global Positioning System (GPS) and assisted GPS reception.

Most of our work has been done using Apple iPad tablets (including iPad mini) and smartphones (iPhone) so these devices, and iOS apps, are mentioned frequently in the book. The book uses these as a model system but hardware and apps from other manufacturers of tablets will work just as well, perhaps even better. We use the commonly accepted term 'apps' for an application running on a tablet. Increasingly, versions are developed for both iOS and Android operating systems. Although we may mention a particular app there will almost certainly be others available. Our advice is to experiment, especially as apps are either free or inexpensive. Free-versions can often be useful to trial the functionality of an app and the subsequently upgraded to obtain more facilities or a wider range of features. Most frequently, the simplest version will work well enough for field use.

In time, we hope that students will start to use their own tablets and smartphones for field use; so the cheaper the system the better. We have not provided ® or ™ signs but merely indicate a proprietary brand by placing it in *italics*. As already stated, there will be many more apps that could provide similar functions and these will continue to be developed and advanced over time.

Table 1.2 A basic consideration of the facilities available on most tablet devices

Sense	Mode	Tablet, basic functionality	Example	Additional functionality	Additional functionality/notes
Audio	One to one, synchronous	Microphone, speaker	Phone chat	External microphone	External (& BT) headset
	One-to-one, asynchronous	Microphone, speaker	Voice memo		
	One to many, synchronous				External (& BT) speaker
Video	One to one, synchronous	Forward facing camera	Video chat		
	One-to-one, asynchronous		Video message		
	One to many, synchronous		Skype		
Still images	One to one, synchronous	Camera	Photograph	Supplementary lens	BT microscope
Still images	One to many	Interactive visualisation	Word cloud generator		Creation on desktop
Text reading	One to one	Download text	E-book		
Location	Instantaneous	GPS, aGPS	Map location	External GPS dongle	
Position	Instantaneous	Accelerometers	Clinometer		
Writing	Instantaneous	On screen	Photograph annotation	Capacitive pen	
Typing	Instantaneous	On screen typing		Capacitive pen	External (BT) keyboard
Drawing	Instantaneous	On screen	Field sketch		
Recording	Screen, one to many, asynchronous	Camera	OCR of text		
Recording	Screen writing	On screen	Note-taking		

(continued)

Table 1.2 (continued)

Sense	Mode	Tablet, basic functionality	Example	Additional functionality	Additional functionality/notes
Recording	Screen writing	May require internet access	Handwriting recognition		Active external pen device
Recording	Instantaneous	Microphone	Voice recognition		
Viewing (+ audio)	Synchronous, one to many	BT or HDMI connector	Video projection to TV	BT and hardware links to external devices	WiFi link to TV or data projector
Internet	Instantaneous	With browser	Surfing	Requires internet connection	
Vibration	Instantaneous-continuous	Accelerometers	Seismic activity		
Sound	Instantaneous-continuous	Audio input	dBA levels		
Light	Instantaneous-continuous	Camera input	Light levels for photography		
Temperature	Instantaneous-continuous	Via plug in sensor	Changing room temperature	External sensor required	
Storing data	Internet and ex-internet connected	Wi-Fi inbuilt	Cloud storage	External USB-Wi-Fi dongle	
Using data	Instantaneous	As computer	Data analysis	Manual, CSV file input	

1.10 Summary

This book suggests ways of using technology (including mobile devices) to enhance student learning; we suggest you consider the basic SETT requirements (see Zabala 2005 for more detail) to develop what you want students to do in their activities and how this can be supported with the most appropriate technology.

> **SETT Framework for student learning**
> **Student**—those students on the module needing to experience 'fieldwork' at an appropriate level; what do they need to do; what are their abilities, interests, concerns etc.
> **Environment**—an appropriate field environment determined by course conditions, location, duration of stay etc.
> **Task**—experiential fieldwork activities that are designed to deepen students' understanding of their subject and/or develop new skills
> **Tools**—equipment, devices, services and strategies to support the above

In Chap. 2 we start with a consideration of what tablets can do, or rather, what you might want to do with tablets to support your field teaching. By seeing what others have done, via the case studies in the subsequent Chapters, we hope that you will think out good ideas for your own fieldwork with students in progressing from novice to expert. Although the technology used in the case studies is designed to enhance student learning in fieldwork it is not necessarily complex and may just use simple facilities in new or different ways.

By virtue of the individual treatment of the examples given we have only provided a general framework and we have left tutors to tell their own story and in some cases to show student involvement. With the passage of time staff and modules change and not all the case studies show existing practice. All the examples show what Technology Enhanced Learning can do for your fieldwork activities.

> This book reflects a particular point in time, but the supporting website http://www.enhancingfieldwork.org.uk continues to evolve. Do check this online resource for updates, new case studies and the latest developments in technology-enhanced learning for fieldwork.

References

Andrews J, Kneale P, Sougnez W, Stewart M, Stott T (2003) Carrying out pedagogic research into the constructive alignment of fieldwork. Planet Spec Ed 5:51–52. Retrieved from http://78.158.56.101/archive/gees/pubs/planet/pse5back3.pdf

Beddall-Hill NL, Jabbar A, Al Shehri S (2011) Social mobile devices as tools for qualitative research in education: iPhones and iPads in ethnography, interviewing and design-based research. J Res Cent Educ Technol 7(1):67–89

Beetham H (2013) Designing for active learning in technology-rich contexts. In: Beetham H, Sharpe R (eds) Rethinking pedagogy for a digital age. Routledge, New York, pp 31–48

Campbell SW (2006) Perceptions of mobile phones in college classrooms: ringing, cheating and classroom policies. Commun Educ 55(3):280–294

Caudill JG (2010) A futurist perspective on mobile learning. In: Guy R (ed) Mobile learning: pilot projects and initiatives. Informing Science Press, Santa Rosa, pp 253–271

Christensen CM, Horn MB, Johnson CW (2008) Disrupting class: how disruptive innovation will change the way the world learns. McGraw-Hill, New York

Cotton DRE, Cotton PA (2009) Field biology experiences of undergraduate students: the impact of novelty space. J Biol Educ 43(4):169–174

Cochrane T, Bateman R (2010) Smartphones give you wings: pedagogical affordances of mobile Web 2.0. Australas J Education Tech 26(1):1–14

Easton E, Gilburn A (2012) The field course effect: gains in cognitive learning in undergraduate biology students following a field course. J Biol Educ 46(1):29–35

France D, Ribchester C (2004) Producing websites for assessment: a case study from a level 1 fieldwork module. J Geogr High Educ 28:49–63

France D, Whalley WB, Mauchline A (2013) Using mobile devices to enhance undergraduate field research. CUR Q 34(2):38–42. Council on Undergraduate Research, Washington, DC

Fuller IC, Edmondson S, France D, Higgitt D, Ratinen I (2006) International perspectives on the effectiveness of geography fieldwork for learning. J Geogr High Educ 30:89–101

Gamarra JGP, Ironside JE, de Vere N, Allainguillaume J, Wilkinson MJ (2010) Research-based residential fieldwork learning: double bonus? Biosci Educ 16(6). Available at http://journals.heacademy.ac.uk/doi/pdf/10.3108/beej.16.6

Goulder R, Scott GW, Scott LJ (2012) Students' perceptions of biology fieldwork: the example of students undertaking a preliminary year at a UK university. Int J Sci Educ 1:22

Guy R (2010) Mobile learning defined. In: Guy R (ed) Mobile learning: pilot projects and initiatives. Informing Science Press, Santa Rosa, pp 1–7

Healey M, Jenkins A (2000) Kolb's experiential learning theory and its application in geography in higher education. J Geogr 99:185–195

Hovorka AJ, Wolf PA (2009) Activating the classroom: geographical fieldwork as pedagogical practice. J Geogr High Educ 33:89–102

ipadsinHE (2014) International conference on the use of iPads in higher education. Retrieved from http://www.ipadsinhe.org

Jarvis C, Dickie J (2010) Podcasts in support of experiential field learning. J Geogr High Educ 34:173–186

JISC (2014) www.jiscdesignstudio.pbworks.com

Johnson L, Adams Becker S, Cummins M, Estrada V, Freeman A, Ludgate H (2013) NMC horizon report 2013: higher education edition. The New Media Consortium, Austin

Katz JE (2005) Mobile phones in educational settings. In: Nyiri K (ed) A sense of place: the global and the local in mobile communication. Passagen, Vienna, pp 305–317

Keeton MT, Tate PJ (eds) (1978) Learning by experience—what, why, how. Jossey-Bass, San Francisco

Kern EL, Carpenter JR (1984) Enhancement of student values, interests and attitudes in Earth Science through a field-oriented approach. J Geol Educ 32(5)

References

Kern EL, Carpenter JR (1986) Effect of field activities on student learning. J Geol Educ 34:180–183

Knight J (2006) Investigating geography undergraduates' attitudes to teaching, learning, and technology. Planet 19:19–21

Kolb DA (1984) Experiential learning: experience as the source of learning and development. Prentice-Hall, Englewood Cliffs

Koole ML (2009) A model for framing mobile learning. In: Ally M (ed) Mobile learning: transforming the delivery of education and training. AU Press, Canada, pp 25–47

Mac Callum K, Jeffrey L (2010) Resistance to the inclusion of mobile tools in the classroom: the impact of attitudes and variables on the adoption of mobile learning. In: Guy R (ed) Mobile learning: pilot projects and initiatives. Informing Science Press, Santa Rosa, pp 9–26

Martin F, Ertzberger J (2013) Here and now mobile learning: an experimental study on the use of mobile technology. Comput Educ 68:76–85

Maskall J, Stokes A (2008) Designing effective fieldwork for the environmental and natural sciences. In: GEES Teaching and Learning Guide. HE Academy Subject Centre for Geography, Earth and Environmental Sciences, Plymouth

Masrom M, Ismail Z (2010) Benefits and barriers to the use of mobile learning in education: review of literature. In: Guy R (ed) Mobile learning: pilot projects and initiatives. Informing Science Press, Santa Rosa, pp 9–26

Mehigan TJ, Pitt I (2010) Towards an ubiquitous future: modeling existing mobile learning system research. In: Guy R (ed) Mobile learning: pilot projects and initiatives. Informing Science Press, Santa Rosa, pp 273–290

Melhuish M, Falloon G (2010) Looking to the future: m-learning with the iPad. Comput NZ Schools Learn Lend Technol 22(3):1–16

Parr DM, Trexler CJ (2011) Students' experiential learning and use of student farms in sustainable agriculture education. J Nat Res Life Sci Educ 40:172–180

Peacock J, Park JR, Mauchline AL (2011) Effective learning in the life sciences: fieldwork. In: Adams D (ed) Effective learning in the life sciences. Wiley, Chichester, pp 65–90. ISBN 9780470661567

Prokop P, Tuncer G, Kvasničák R (2007) Short-term effects of field programme on students' knowledge and attitude toward biology: a Slovak experience. J Sci Educ Technol 16(3):247–255

Rickinson M, Dillon J, Teamey K, Morris M, Choi MY, Sanders D, Benefield P (2004) A review of research on outdoor learning, vol 87. FSC occasional publication

Scott GW, Goulder R, Wheeler P, Scott LJ, Tobin ML, Marsham S (2012) The value of fieldwork in life and environmental sciences in the context of higher education: a case study in learning about biodiversity. J Sci Educ Technol 21:11–21

Shih Y, Mills D (2007) Setting the new standard with mobile computing in online learning. Int Rev Res Open Distance Learn 8(2):17

Swansborough S, Turner D, Lynch K (2007) Active learning approaches to develop skills for sustainability. In: Roberts C, Roberts J (eds) Greener by degrees: exploring sustainability through higher education curriculyes, afraid sola. Geography Discipline Network, University of Gloucestershire, Cheltenham

Stokes A, Boyle AP (2009) The undergraduate geoscience fieldwork experience: influencing factors and implications for learning. In: Whitmeyer SJ, Mogk DW, Pyle EJ (eds) Field geology education: historical perspectives and modern approaches, vol. 461. Geological SocAmer Inc, Boulder, pp 291–311

Traxler J, Wishart J (eds) (2011) Making mobile learning work: case studies of practice. ESCalate HEA Subject Centre for Education, Bristol

Wei R, Leung L (1999) Blurring public and private behaviors in public space: policy challenges in the use of improper use of the cell phone. Telematics Inform 16:11–26

Weiser M (1991) The computer for the 21st century. Sci Am 265(3):66–75

Welsh KE, Mauchline AL, Park JR, Whalley WB, France D (2013) Enhancing fieldwork learning with technology: practitioner's perspectives. J Geogr High Educ 37(3):1–17

Whalley BW, France D, Park JR, Mauchline AL, Powell V, Welsh K (2014) iPad use in fieldwork: formal and informal use to enhance pedagogical practice in a bring your own technology world. In: Souleles N, Pillar C (eds) Proceedings of the first international conference on the use of iPads in higher education, Paphos. www.ipadsinhe.org. 20–22 Mar 2014. ISBN: 978-9963-697-10-6

Wheeler A, Young C, Oliver K, Smith J (2011) Study skills enhancement through geography and environmental fieldwork. Planet 24:14–20

Woodcock B, Middleton A, Nortcliffe A (2012) Considering the smartphone learner: an investigation into student interest in the use of personal technology to enhance their learning. Stud Engagem Exp J 1(1):1–15

Zabala JS (2005) Ready, SETT, go! getting started with the SETT framework. Closing Gap 23 (6):1–3

Chapter 2
Introduction to Tablets and Their Capabilities

Abstract In this chapter we provide a few introductory notes on using and optimising tablets for fieldwork learning. Tablets come with few instructions, however some users are not used to exploring the operating system or realise that you can achieve a result in a number of different ways. Supporting documents are available e.g. for iPads at Apple (2014) as well as number of basic guides, see Turner (2014). While this chapter is not a 'how to' manual it seems appropriate to provide some extended notes relating to the fieldwork functionality of tablets with regards for example, connectivity, accessibility and data storage.

Keywords Tablet computer · iPad · Field-ready · Functions · Internet connection

2.1 Usability of Tablets

The usability of tablets depends upon the size and, to some extent, price. Thus, although smartphones fulfil many of the capabilities in Table 1.2 they provide rather small screens, even at high resolution, and relatively poor battery life compared with the larger tablets. Battery life is continually improving but 10 hours use of a tablet should be achievable even with power intensive uses such as GPS location. We used iPad 2s from the start on our project but the smaller iPad mini has meant increased usability in the field because it fits into a pocket more easily.

Students generally described iPads as 'easy' and 'useful' when answering a usability questionnaire after a field session using iPads to support their learning. Figure 2.1 shows a 'word cloud' of a typical class response to the use of iPads after these two main responses have been removed.

When people ask about using tablets as notebooks in the field the question immediately raised is 'what do you do when it rains?' The cheapest solution is to use a polythene bag with a few grains of rice to absorb any moisture. It works; as does a waterproof housing for underwater use in extreme fieldwork such as marine archaeology or a shock-proof, waterproof case for general outdoor use (Fig. 2.2).

© The Author(s) 2015
D. France et al., *Enhancing Fieldwork Learning Using Mobile Technologies*,
SpringerBriefs in Ecology, DOI 10.1007/978-3-319-20967-8_2

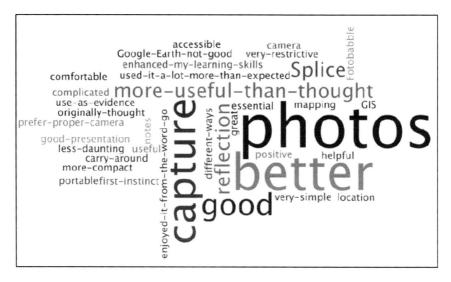

Fig. 2.1 A word cloud bringing together the main comments from a questionnaire after 'easy' and 'useful' had been deleted

An immediate advantage of using a ruggedised tablet as both a field guide as well as a notebook is that it does avoid juggling a paper notebook, pencil and identification guide when working in a bog in the rain.

For our students' use of the project's iPads we have supplied them in a ruggedised case as shown in Fig. 2.2. These cases work well to protect the equipment

Fig. 2.2 iPad Mini in Griffin 'Survivor' case

2.1 Usability of Tablets

Fig. 2.3 Word cloud formed from students responding to 'disadvantages' in a post field trip questionnaire on the use of iPads in the field

from the weather and from knocks but are inconvenient when access is needed to the charging port or camera/microphone connections. They also make the device much bulkier. For most purposes a simple case with a screen cover works well enough. For class use however, ruggedised covers are probably a worthwhile investment as students' main concerns over the use of iPads in the field reflect worries over damaging expensive department-owned equipment (Fig. 2.3).

One word not used in our survey is 'glare' or 'reflection' or associated problems from direct bright sunlight. This may be a consequence of fieldwork carried out in the British Isles! Several types of proprietary anti-glare (and screen protector) kits are available. However, it has to be admitted that glare may be a problem and this is where a high contrast screen (such as on the Kindle e-book reader) performs well.

Fig. 2.4 iPad Mini with BT keyboard acting as a rest. Below is the holding part of a case with the screen flap turned back and key ring holding an extending safety line. To the *left* is an Apple BT full size keyboard, BT earphones and microphone and the silver 'drum' is a desk-top BT loudspeaker

(Note that these high contrast screens are sometimes called 'electronic paper', 'e-paper' or 'electronic ink' screens but these terms cover several types of technology.) At present, this disadvantage with iPads needs to be weighed against their much more general usability compared with e-book readers.

Although much data entry, from writing to entering values on a spreadsheet, can be done via the screen keyboard this can become tedious and difficult for more than a small quantity or time. Some tablets come complete with external keyboards (such as the Microsoft 'Surface') others need them bought as extras. These will usually fit snugly but a full-size Bluetooth (BT) keyboard could be used in the office. Made to measure external keyboards are still a little cramped for use on an iPad Mini. Figure 2.4 shows relative sizes of some devices and Bluetooth keyboards. In practice you can use a conventional keyboard at home for protracted typing and the small device in the field for notes.

2.2 Getting Used to a Device

We suggest that tutors, as much as students, become familiar with their devices before using them in the field. Although our students are supposed to be 'digital natives' this ability may only go as far as using social media and taking photographs/videos and emailing them. Students should be given opportunities to become familiar with the new technology and be made aware of the means of saving and storing attachments, and we devote some attention to this below. They should also be aware how to use screen gestures, keyboard shortcuts, 'undoing', closing down apps in multi-tasking and basic aspects such as rebooting and taking a screen-shot. To assist with familiarisation there are various on-line guides and help forums as well as guidebooks available on the market.

It will not take long but a little time spent getting used to a device will, or should, ensure more efficient time spent in the field. It is important to check that the equipment has the appropriate apps loaded and is fully charged before going into the field! This is also a good time to think about students with any known disabilities. There are now many apps (and some hardware devices) that will enable disabled students to perform better and not be seen to require special attention. In any case, we suggest that all students are clear on the use of devices and apps before they go into the field. We have several examples of pre-field trip experience in Chap. 7 and these could be used to get students accommodated to the devices and apps to be used in the field.

At the least, we suggest that students (and tutors!) know how to do the following:

- 'undo' instructions from the keyboard
- know how to use keyboard shortcuts for typing
- how to take a screenshot and access it

- monitor battery and memory storage space use
- share/export/send/open files in another app and use *AirDrop*

2.3 Communication

Tablets, like smartphones, are essentially communication devices. Smartphones, like any telephone, tend to be thought of in a traditional manner. Talking, or texting, to another person is this basic. However, Table 1.2 extends the ways that we can use tablets to communicate with students. It tries to provide a basic rationale for using tablets and apps as an enabling technology that is only just now possible. Chapter 6 offers detailed case studies of how communication can be facilitated between staff and students.

2.4 Accessibility

Both Apple and Windows operating systems have gone to great lengths over the years to improve 'accessibility'. That is, allow for people who may be disabled in one way or another or have special needs. In the UK, JISC's TechDis (2014) is an advisory service for inclusion and highlights ways in which computer technology can be used to help accessibility and inclusivity in Higher Education.

One major function of tablets is the ability of tablets to be used to improve accessibility and this often means using the tablet to transfer information from one 'sense' to another, voice recognition for note-taking is one such. The ability of tablets to combine data input and move it between senses (Table 1.2) is one where the inventiveness of app designers has had considerable input to education and where fieldwork, in all its manifestations, benefits. For example, cold fingers often make traditional note-taking with pencil and notebook difficult. A tablet's screen keyboard alleviates this problem but a voice note taker may be the answer. Anything other than a tablet or smartphone means a larger cumbersome laptop is necessary and a dedicated voice record one more device to juggle with.

Apple's *Siri*, provides a measure of on board task management by voice recognition that may be useful. This capability will undoubtedly increase as computing power and sophisticated algorithms improve. If keyboard entry is needed then some apps will make the process easier; one such is '*Fleksy*'. In general, tablet manufacturers have developed ways of accommodating to users' needs. The use of keyboard short-cuts is one of these, although perhaps the most significant is the use of gestures. Note however, that it is not possible to use a Bluetooth mouse or trackpad with iPads. However, a capacitative stylus or pointer is useful, especially for drawing and annotating applications.

2.5 Connectivity

It is not always necessary to have any Wi-Fi connection in the field or even 3/4G mobile Internet coverage. Indeed, in some locations fieldwork may be taking place well beyond their reach. Most apps will work on the tablet and do not rely on computation being done interactively on a website. iPods and other similar devices act as useful, less expensive devices than tablets but lack some functionality as not all apps will work on them. We suspect that these will go out of fashion as tablets with full interconnectivity (i.e. with 3/4G) become cheaper. Kindles and other e-book reader devices can certainly act as field devices but in a limited way. The usability of tablets like the iPad and Galaxy are preferable for field, and general educational, use.

An important distinction between the Wi-Fi only and the Wi-Fi+3G iPad is their ability to receive Global Positioning System (GPS) signals. The Wi-Fi + 3G iPad and the iPhone both have a GPS receiver. The Wi-Fi-only iPad and the iPod do not; these devices require an accessory to receive these signals. GPS signals come directly from the network of GPS satellites and provide latitude and longitude information. This function is of great use in fieldwork as devices with internal GPS receivers can log location information in remote places without Wi-Fi or 3G connections. However, the digital compass and accelerometers are in both WiFi and 3/4G varieties and also the latest iPad Air version. In choosing an iPad therefore, not only are physical size and connectivity important factors to consider but also processor speed and memory available. It is wise to do a model comparison, especially for recommending any bulk purchases for an institution.

2.6 Digital Mapping

Maps provide the basis for many aspects of fieldwork. *GoogleMaps* and *GoogleEarth* are perhaps the best known free base map possibilities. For national surveys, whether of topographic, road or Lidar imagery as well as satellite data, price structures may vary according to current situations. In the UK, Edina (2014) delivers a variety of services including Digimap (which allows free access to maps for subscribing UK universities and colleges). In the USA, the USGS and NASA provide similar facilities. Although an eye needs to be kept on reproduction and intellectual property rights (IPR), it is possible to facilitate field mapping by capturing an image via the iPad camera and importing it into an app such as *Skitch*, where it can be annotated (and see more about this app in Chap. 3). Further information concerning geo-referencing/geotagging photographs is provided in Chap. 4, Sect. 4.5 and in Case Study 8. Specialised fieldwork apps that have digital mapping functionality are considered in Chap. 8.

2.7 Data Storage

With 'the Cloud', becoming increasingly present in populated areas backup and data transfer is now relatively easy to set up. Tablet manufacturers have their own systems; Apple (*iCloud*), Android (*Google Drive*) and Microsoft (*SkyDrive*) devices also have a pick of several sites to which backups can be made and files shared between users. See Sect. 2.8 on the use of *Dropbox* as a generic system for data storage.

There are other 'cloud' storage systems available such as *SugarSync*, *Copy* and *Zipcloud*. A web search will show what is currently available as well as their costs and storage capacities.

There are also flash memory devices that can act as local Wi-Fi-storage. Such devices (such as Transcend's *StoreJet* and the *AirStash*) allow you to provide access to several tablet devices and save memory by carrying video and music off the tablet. These are useful for data sharing and backup if cloud storage is not available.

As well as normal storage 'on board' storage there are of course image-dedicated storage and display systems of which *Facebook* and *Picassa* are the best known. Storage limits are variable and for heavy usage it may be better to subscribe to a cloud-based storage site. There are also social networking apps to share images and video. *Instagram* is one such and *YouTube* and *iTunes U* are widely used in an educational context. As professional photographers now use iPads on shoots there are now cloud-based image storage apps available for such 'high end' use. Although there are cost and size limitations of these facilities we leave it to the user to choose what seems best for their purposes.

2.8 Data Sharing

It may well be that students have to share a device in a small group, not only because of cost but because group work is common in the field. Students might need to share information with staff or other groups. Some apps, such as *Bump*, avoid the use of iTunes, others (e.g. *AirForShare*) allow file sharing between different OS devices. The lack of a USB 'stick' might seem a disadvantage but there are several apps that can be used to share information locally. *Airstash*, as just mentioned, is one such storage device with Wi-Fi.

> *Dropbox* is a free piece of software that can be used on a desktop/laptop, Smartphone or on a tablet computer such as an iPad or a Samsung Galaxy Nexus. Dropbox folders can be shared across class groups, offering an alternative to the VLE during fieldwork. *Dropbox* is a cloud storage system and syncs automatically, which means any changes made to a document on a

desktop computer are automatically updated on the tablet or Smartphone version (and vice versa).

One of the main benefits of *Dropbox* is that it is an easy way to upload documents from a Smartphone/tablet to the desktop version and does not require iTunes. Folders can be shared which is ideal for collaborative work in groups, which takes place on many field courses. Users have access to 2 GB of space for free; more space can be purchased if necessary. It is regularly cited as being a "must have" app for the iPad, which at present has no inbuilt file organisation/storage system. However, there are file management apps (such as *Documents*5) that can interact with *Dropbox*, *Google Drive* etc. Dropbox is available for Windows, iOS, Kindle Fire, Blackberry and Android platforms. For further details see Dropbox (2014).

Field data can be collected and shared using on tablets through *FieldtripGB*. This app created by EDINA at the University of Edinburgh can be used offline and uses *Dropbox* to upload and synchronise the data when in a Wi-Fi hotspot. A bespoke field data collection form can be created, prior to the fieldwork and subsequently presented on a base map or exported for further analysis. Other field trip notebooks are available and discussed in Chap. 4.

Showing images, results, and data to groups is often an important part of fieldwork sharing and feedback provision. With an iPad or iPhone you can do this in several ways. If there is a video projector with a VGA input (such as you would normally plug into a laptop) then a Apple 'Lightning' connector to VGA allows you to do this easily. Most projectors pick up the signal easily.

A local wireless network can be established between the iPad and the video projector with the aid of WPS2 Dongle. Apps such as *Wifi-doc* or *MobShow Lite* will display presentations, images and PDFs via Wireless Dongle to enable any mobile device to share the VGA projector. Alternatively you can use AppleTV to wirelessly connect you iPad. An easy way to hook up to most modern LCD/plasma screens with HDMI outputs is by using an HDMI-HDMI cable (being careful to note the sizes of terminal) and suitable connector to the iPad.

Providing amplified audio output from an iPod or even iPhone is easy with either small, unamplified speakers or, rather better, a Bluetooth-connected speaker (e.g. HMDX Jam Plus) with internal battery.

2.9 Data Processing

Although students tend to use a spreadsheet to process data back at University, this might not be the best way to proceed. Apps are now available (*Numbers*, *Excel*) for students to have the ability to collect data and process data in the field.

It may be however that a 'standard' calculator might be the best solution for occasional use. A wide variety is now available, including graphing calculators. A simple notepad calculator, *MyScript*, on which you write with a finger or stylus on the screen is editable and fast, a useful app and good for teaching maths and as a field calculator.

For analysis of pairs (x, y) of data the scientific *DataAnalysis* app provides a wide variety of curve fits and basic analysis, as well as export options. However, it does yet not accept a csv file, although this may not be a major problem.

The increasingly popular programming language Python has an iPad shell *Python Math* for command-line data and there are apps using this for a variety of tasks. For Unix aficionados, *Vim* is available along with other tools that are probably beyond the needs of most fieldworkers. They do however show the ingenuity of programmers in turning a 'toy' into a scientific tool.

2.10 Data Collection and Loggers

As mentioned previously, the on-board sensors (accelerometers) in iPads have allowed several apps and features suitable for fieldwork. These include the seismometer (e.g. *iSeismometer* see Takeuchi and Kennelly (2010) for laboratory demonstrations), the clinometer which is utilised in some of the ranging and survey apps mentioned above and for geological applications such as lambert which are mentioned in Chap. 8.

EasySense is an app that takes data from a variety of sensor devices, mainly for laboratory use.

There are, to date, relatively few plug in sensors for iPads. A useful one, at some expense however as it has the temperature sensor (PRT) as a physical plug in is *iCelsius* (see Chap. 8).

iBeacon technology (see Thompson 2013) enables a device to transmit its position to another device e.g. Smartphone or iPad, through Bluetooth technology. This has the potential to be become widely adopted. The Department of Geography, University of Otago, New Zealand have trailed this technology by attaching *iBeacons* to river gravels to monitor and trace their movement after storm and flood events.

Colleagues (Trevor Collins and Sarah Davies) at the Open University, UK, are exploring news ways to record real-time data through relatively inexpensive Bluetooth enabled devices such as *SensorTag*, *SensorBug* and *Wimoto*. Any Android or IOS device can connect wirelessly to these low-powered sensors, to interrogate and download the stored data.

2.11 Summary

This chapter has listed the various functions of tablets and Table 1.2 describes the wide variety of ways in which you may wish to use them. We encourage you to think of imaginative ways of doing things to enhance your modules and fieldwork by using this information and the schemes in Figs. 1.1–1.3. We provide case studies in the following chapters to illustrate technology use under educational situations. In some cases we have just made suggestions of how particular facilities (such as audio devices) and apps might be used.

We would like to stress a likely difference between you, as tutor, and your students. We hope you become familiar with the range of things that can be done with tablets and browse websites for other examples. However, while your students may be familiar with the use of tablets and/or smartphones they may not be familiar with using them to support their learning. Students will almost certainly need to be shown what can be done and the wide range of apps that are available. We have tried to give instances in this book. Over time, the concept of Bring Your Own Device/Technology (BYOD/T) is becoming increasingly well known within education as a means for students to use their own devices within formal education activities.

Having set the educational scene we now look at ways in which the items in Table 1.2 can be introduced into student activities.

References

Apple (2014) Supporting OS manuals. Retrieved from http://support.apple.com/manuals/
Dropbox (2014) Cloud storage website. Retrieved from http://www.dropbox.com
Edina (2014) Online services for UK higher and further education and beyond. Retrieved from http://edina.ac.uk
Takeuchi K, Kennelly PJ (2010) iSeismometer: a geoscientific iPhone application. Comput Geosci 36(4):573–575
TechDis (2014) UK advisory service on technologies for accessibility and inclusion for disabled staff and students. Retrieved from https://www.jisc.ac.uk/
Thompson D (2013) Beekn: beacons, brands and cultures on the internet of things. Retrieved from http://beekn.net/guide-to-ibeacons/
Turner P (2014) Teaching geography in a digital world. Retrieved from iBooks https://itun.es/gb/h2TYZ.l

Chapter 3
Capturing and Using Visual Imagery in the Field

Abstract This chapter provides a discussion, examples and case studies mainly associated with visual imagery. We provide a general illustration of what you can do with mobile technologies and how they have changed, and are changing, tutors' views of education. More detailed aspects of photography and related apps are covered in Chap. 8.

Keywords Tablet computer · ipad · Digital imagery · Digital camera · Visualisation

3.1 Still Visual Imagery on Tablets

Visually-sensed information (Table 1.2) is perhaps the most intuitive and basic use of tablets; viewing a website or taking a photograph for example. The former requires a synchronous Wi-Fi access, as would a laptop, so it may be necessary to plan student use before using tablets in the field. In our experience, locations in hotels or study centres with Wi-Fi coverage may have Wi-Fi networks that wilt under the stress of students trying to access and download large files. It may be a good idea to download materials in advance of a visit to a centre. Note too that networks may suffer if many students are trying to access a cloud-based application.

3.1.1 Examples of Use—The Basics

You might want to get students to download and store images for use or recognition in the field. This might be at any level of complexity; a reminder of an architectural style such as a Norman arch or ionic order capital is a simple example. Students would need to download the image, perhaps itself the result of a web search before going into the field, and then store the image as a reminder for use in the field.

The image, probably along with others, could then be built into a field notebook or guide (see Sect. 4.4 for additional information on digital field notebooks). Basing such a task on student interactivity, getting them to find the appropriate image, rather than the tutor producing a guide *for* them is advantageous for learning. It also provides an opportunity for students to think about the use of intellectual property rights (IPR), searching websites in a meaningful way as well as providing metadata about the images used.

The simplest way of using these images would be to store them in the device's photo library. This is simple, but perhaps confusing if there are more than a few. Alternatively, the images could be brought together as a simple collage (e.g. *PhotoGrid*) or slide show. Even better would be to place them, with appropriate annotation and space for notes, in a *Pages* or *Keynote* document. These can be shared in a group and PDFs made directly from them at the exporting stage. Many educational variants can be used in this way, from quizzes to 'iSpy' or filling in gaps as a form of 'treasure hunt' at any level of sophistication. Geocaching is a variant of an old-fashioned treasure hunt that can be brought in with digital technologies.

Another traditional approach would be to provide a full field identification guide for an activity. This might be web-based or delivered in PDF format for downloading and subsequent examination in the field. Many households will have paper-based bird identification guides. We now have tablet-based guides that will do the same job and even provide an audio recording of the call or song of the bird as well as the ability to take notes. The price in 2014 seems to be about the same for a good bird identification app as for a book. To make this useful in the field, all the materials are downloaded to one iPad. So field guides can be constructed digitally and we look at this in more detail in Case Study 22.

For advanced work however, where a taxonomic (binomial system) key is required for field use, a published field guide may be uneconomic to produce as a commercial e-book or e-guide. Furthermore, one might not expect all students to buy one. Two possibilities might be considered; to take the field guide, but be prepared for bad weather, and students take photographs of 'unknown' examples for future reference. The tablet computer could be used to take photographs of plants, rocks and architecture but be rather less informative if the subject was a bird! We give a more detailed analysis of the apps available for field identification at E-fieldwork (2014).

An alternative approach would be to produce your own guide from down-loaded images and tables. As more subjects go digital it becomes easier to obtain good quality material, often copyright free, for academic institutions. This is a good opportunity to get students to be involved actively. Rather than a Cooke's Tour approach, 'at this outcrop you will see ….', students can do some research before the day and perhaps explain or discuss the feature to each other in the field. These are aspects of pre-field trip preparation that we shall pick up in a later Chap. 7.

We now present some case studies and examples showing a variety of ways in which visual images can be used by students in a variety of active ways, some very simple, others more complex. The first case study is a simple way in which Andrew

Goodliffe has used tablet technology to enhance the learning experience for his students. At its simplest, even a single device can be used but Andrew shows how sharing materials can be made to work.

Case Study 1: How the iPad Transformed My Teaching (Andrew Goodliffe)

Institution: University of Alabama, USA

Keywords: *Dropbox, GIS, Google Earth*

Aim: To show some general ways in which tablet technology can be used.

The iPad can easily be used in classrooms and, due to its size and ease of use, it can even be passed around the class so that students can view documents or photos. In the future, multiple iPads could be used in classroom situations so that students can use them in small groups or individually. (Note, we have suggested other ways of sharing materials in Chap. 2).

The majority of a teacher's photos and scientific papers can be stored on an iPad. There are a number of filing systems, such as *Air Sharing* and *GoodReader*, available to use in class. In a teaching environment, the best system for the iPad could be *Dropbox*, which became our default file and share system. Files can be instantly uploaded and shared with students and teachers.

Using the 3G connection the iPad can use *Google Earth* and other GIS applications to collect data and it can be used as a virtual map. Students can use the iPad to see exactly where they are in real-time. The iPad allows students to carry out their work in the field and analyse it on-site. They are also able to upload their data and share it with other students and teachers. This gives the students a greater sense of understanding and ownership of geophysical field data.

Technology and apps required:
Dropbox and GIS applications such as *Google Earth*.

The iPad can be used wherever you are in a number of situations, as it is a mobile technology. Files can be shared with students in formal and informal settings. The iPad can also be used in fieldwork to use various applications including GIS technology.

Pitfalls/Problems/Limitations:
A possible limitation with the iPad is the lack of an obvious file system. However, there are various file sharing systems, such as *Dropbox* and other ways of sharing have been mentioned previously.

Technological expertise required:
Staff skills required: Must be able to use the iPad and the applications that are used on it.

Student skills required: Students must be able to use the iPads and the various applications that are to be used on it.

Also see: Goodliffe (2011)

The next case study, by Ian Stimpson, is a reflection on mobile platforms and how they can be used to enhance the learning experience for students in geography and geology. This applies however to many aspects of 'out of classroom' activities and not just 'fieldwork' in the traditional sense. Note that this project was done before iPads but that iPods could (and still can) fulfil the roles of active field education. In particular, one of the main objectives 'was to evaluate the media devices vs. traditional paper for learning'. This case study makes valuable reading in an iPad world.

Case Study 2: E-learning or a-Gimmick? Evaluating the Use of Rich Media in Geography Fieldwork (Ian Stimpson)

Institution: University of Keele, UK

Keywords: E-learning; mobile; play station, rich media, android

Aim: What mobile platform is best at supporting rich media for student fieldwork learning?

Many universities now recognise that multimedia platforms—such as mobile/smart phones and mp3 players—are part of everyday student life and hold great potential to enhance fieldwork learning. As a response, a technology-focused project was set up to answer a number of questions related to the usefulness of cross-platform rich media resources in geographical fieldwork pedagogy.

The two main questions were:

- Does the technology distract or enhance the student fieldwork learning experience; does it make them more or less aware of their physical surroundings?
- Which is better: Digital media versus Paper hand-outs? Is it e-Learning or just a-Gimmick?

To tackle these questions the project team created an audio-visual tour of Bath, which was to be tested using a focus group of undergraduate student volunteers; evaluating the usability, suitability and effectiveness of such a tour on different media platforms. In conjunction with the student focus group, lecturing staff across the GEES community were asked to trial and evaluate such a project to gain an insight into the advantages and disadvantages of using multimedia devices in this context.

3.1 Still Visual Imagery on Tablets

The selected devices for this project were;

- Samsung D900i and Sony Ericsson W800i (Android Mobile Phones)
- Sony PlayStation Portable [PSP] (Handheld games console)
- Apple iPod touch and Apple iPod-5th generation (iOS Media Players)

The devices were chosen for their commercial success and the frequency of personal ownership amongst University of Exeter and Bath Spa University students. Another contributing factor was the wide range of audio-visual display formats available across the device platforms.

The "Podtour" of Bath was created to give a series of waypoints to the student user. As the students walked to each waypoint an integrated package of audio and visual materials would be provided on their devices for that location; usually in the form of imagery and videos with a specific recorded voiceover. This was created through well-known software, which is readily available, and user friendly e.g. *'GarageBand'* and *'iMovie'*.

As one of the main objectives of this project was to evaluate the media devices vs. traditional paper for learning, the students were allowed to use the devices alongside the traditional paper based worksheet so that an effective comparison could be made. The students were given a quick briefing of the exercise aims, making the students aware of the key questions they should consider while undertaking this exercise:

1. How easy is the (hardware and software) to use? (Specifically the ease of locating and navigating through materials)
2. The functionality and practicality of the viewing experience. (Screen size, resolution, effectiveness under different light conditions)
3. How good was the audio quality; are headphones or built-in speakers better?
4. Overall practicality in terms of the task, e.g. how long did the battery last?
5. How effective was it at being an input device i.e. note taking?

Technology required:

- A range of Android Smartphones can be used
- Sony PlayStation Portable [PSP] (Handheld games console)
- Apple iPod touch and Apple iPod-5th generation (Media Players)
- Computer software such as GarageBand' and 'iMovie'
- Software such as PSPWare, iTunes Transcoding software

Approximate cost of technology:

- Mobile phone prices vary.
- iMovie'—£85 (at the time of use, now £3.99)
- *GarageBand*—range of different types of this software from £15 to £72 (at the time of use, now £3.99)
- *iTunes*—free

What evidence is there of the effectiveness of this activity?
Overall, the devices performed successfully with positive feedback from the students, who clearly identified critical problems with each of the devices.

- Mobile Phone—Students were satisfied with the devices; it offered the familiarity of common mobile phone hardware and navigational tools.
- PlayStation Portable (PSP)—Students showed considerable enthusiasm. Using this gaming device as an educational tool intrigued the students and engaged them. Beyond this, the PSP was a fantastic device that was well suited to the task at hand, showing rich and crisp audio-visual imagery on the bright, wide screen, which offered high resolution detail. Navigating through the system required little explanation before the students became fully conversant with the device. The provision of an optional camera, GPS unit and videoconferencing facilities through recently announced/released upgrades do add appeal to the device.
- iPod—This touch screen device was the best device for in the field functionality, with a clear, large, high resolution screen. Its functionality and practicality were major advantages. This powerful handheld device, with its simplistic user interface, requires no additional explanation from staff. The display automatically readjusts to new aspect ratios, thanks to the inbuilt accelerometers that detects the orientation of the device. Another innovative feature that developed with the iPod is the touch screen 'gestures'.

Students and staff agreed that the PSP was most effective as a replacement for the traditional pen and paper, but traditional, low-cost, tried and tested methods of fieldwork should remain important parts of the student lives.

Pitfalls/Problems/Limitations:

- Mobile phone—Although the mobile phone was familiar to all students there were some flaws to it as a device suitable for fieldwork. Each device had its own software that required different formats, which slowed down the process of uploading, and background knowledge of the device was often required. The screen was too small and hard to see in bright light conditions. Also, the mobile phone's memory is adapted for text messaging and often the large video files slow down the device.
- PlayStation Portable—Performed well in the test, but had a few issues. Although it is a sleek device, compared to the mobile phones and the iPods it appeared bulky to accommodate its large wide screen. The built-in speakers of the device were never designed to be used in crowded busy spaces; they are too quiet and require headphones to listen to the audio clearly. This is not good for group work as the headphones effectively seal off the listener from the outside world. The device's battery life is fair, but far from great, lasting just 1.5 h during the test. This may become laborious to recharge or to have the added weight of spare batteries to carry

round (which cost around £8 RRP). The final problem was students felt unsafe using such a high end, fashionable device so openly in public. For development purposes it is not as easy to update as an iPod, but easier than a phone. PSP has its own video format; Sony's PSPWare software automatically changes the format of your video to suit the PSP making it much more user friendly and time friendly for staff and students.

- iPod Touch and 5th Generation—The 5th generation's screen was adequate but a little on the small side, with more complex menus to navigate with the older 'click wheel' system. However, the iPod Touch showed only one or two negatives, the device worked remarkably well in the test. The notes application that is as standard with this device is not as well integrated as the students would like. They had to drop out of the media player to make notes, which is far from ideal.
- Paper handouts—Students found paper handouts to be not as engaging as the devices.

Another consideration is that a member of staff must be allocated the responsibility for updates and upgrades of the device's software. It is also a limitation of these devices as a whole that they do not stay new for long and in a short timeframe will become obsolete, leading to new costs in the future.

Technological expertise required:
Staff skills required: Needs good understanding of computers to record video and use the software; however once mastered it becomes very easy.

Student skills required: Depends on the device; however most students should cope well with the user-friendly interfaces.

Based on feedback from staff and students, a series of tutorials have been produced to guide practitioners through the simple process of producing rich media resources for MDMDs and can be found at GEES-projects (2006).

We now move on to some more specific ways in which images can be used or enhanced on fieldwork. Again remembering that 'fieldwork' can mean different things to different people.

3.2 Drawing and Painting Tools

The artist David Hockney showed that it is possible to create works of art on an iPad. Many others have done the same with simulated pencils, crayons, pens and brushes. In fact, most of the effects that graphic artists have used on desktop computers now have some form of version on tablets. We would go as far as to say that every tablet used in the field should have a drawing app. Many of these are free,

Fig. 3.1 *Left* field sketch using *Skitch* on its own, *right photograph* with annotation by *Skitch* done in the field and used as a basis for the sketch

although probably with a limited range of tools. There are many to choose from but *SketchBook Express* provides a good start.

One app that we have found very useful (and is mentioned in the following case study) is *Skitch*. This is available for tablet platforms as well as desktops. It can be used for sketching in its own right but can be used to annotate maps and photographs. Both drawing tools and typing can be used. Images can be taken directly or past images can be imported from the photo library. We have found this useful for field situations where students need to make field sketches. They often lack confidence, perhaps because they do not know what to concentrate on. A good way of providing scaffolding, and thus confidence, is for students to take a photograph of the view and sketch over it (Fig. 3.1). The tutor can quickly view each one, make comments to the student, who can then have another try, add requisite information and so on. This app allows support for a variety of field-based sketching and annotation.

The next case study shows a variety of iPad educational applications used by some Field Studies Council centres. Not only do individual iPad apps play a part but Liz Earley and Maryanne Willis (and several other centre tutors) integrate apps for specific projects. Several of these apps also feature in subsequent case studies. This example gives a flavour of what can be done with a little 'educational imagination'.

Case Study 3: Technology in Active Outdoor Classroom Learning: A Case Study from the Field Studies Council (Liz Earley and Maryanne Wills)

Institution: Field Studies Council (FSC), UK

Keywords: *Skitch, GPS Log, Polldaddy, Educreations, Explain Everything, Splice*

Aim: To embed digital technologies, such as iPads and digital cameras into field teaching

3.2 Drawing and Painting Tools

iPads are currently being used by two FSC field centres; embedded into the teaching practices to align with GCSE and A Level curricula. This case study focuses on the integration of mobile devices into fieldwork learning at two field centres operated by the Field Studies Council (FSC); Rhyd-y-creuau in Conwy and Slapton Ley in Devon. Both centres now use iPads to add to students' learning experiences. Field centre staff aim to use these technologies to engage students with the environment that surrounds them without the devices leading to distraction away from the teaching objectives.

iPads have become firmly embedded in the teaching practices of both Rhyd-y-creuau and Slapton Ley Field Centres. Mobile technologies are being used to strengthen links between the FSC and national curricula, as GCSEs now include reference to Geographical Information Systems (GIS) and some A Level examination boards require students to comment on the use of technology in fieldwork. Regional training events, such as the Enhancing Fieldwork Learning Showcase Events, have also provided opportunities to develop new approaches to using technology, discuss ideas and share successes in the use of mobile technologies, such as iPads and digital cameras.

As well as numerical data recording, the devices are used for a wide range of fieldwork experiences. Four apps are focused on here; *Skitch*, *GPS Log*, *Fotobabble and Polldaddy*.

- *Skitch*—a photo annotation app. Used to construct field sketches by annotating directly onto digital photographs. These can be exported as an image or saved using *Evernote*.
- *GPS Log*—a geo-referencing app. Photographs taken in the field (including those annotated in *Skitch*) can be geo-referenced to the current location. Notes can also be attached ('spiked') to this location. Does not require Wi-Fi assess in the field, as most iPads (those with have an inbuilt GPS. On return to the centre these images can be converted to a kmz. file and then viewed in *Google Earth*.
- *Fotobabble*—a photo audio-annotation app, can record voiceovers and attach them to a photograph; in particular this could be used in conjunction with *Skitch*, but the *Fotobabble* can only be outputted through an embedded video or weblink.
- *Polldaddy*—a questionnaire app. Allows quick and professional surveying using multiple choice and longer answer questions outside of Wi-Fi hotspots. Surveys can be constructed through the *Polldaddy* website and then synced to the iPad for use in the field. Responses are uploaded after the survey is completed, the website then produces graphs to summarise responses.

Using *Skitch* (Figs. 3.2 and 3.3) and *GPS Log* in conjunction with each other encourages students to analyse the environment on location through identification and recording of possible reasons for anomalies in their site data. IPads and digital cameras have lessened the challenge of recording data in

Fig. 3.2 Using Skitch to reflect on recreational spaces (Photograph courtesy of Liz Earley)

drizzly and windy weather, when paper leaflets and notebooks are impractical. Even in extreme conditions in wild North Wales and along the blustery Devon coast students were able to capture the scene. On return to the field centre students can share these observations that play a key role in the critical analysis and the drawing of conclusions from their fieldwork investigations.

Storing large amounts of videos, historical photographs, maps and archive materials on the iPads allows students to access a range of secondary materials while at their site location; deepening their understanding of that environment, making links between data sets and looking at changes over time. Using the *Educreations* and *Explain Everything* apps students then produce short presentations in the field using a mix of primary and secondary information to answer questions about the location they are in.

3.2 Drawing and Painting Tools

Fig. 3.3 Part of a *Skitch* to reflect on stream characteristics pinned to a Google Earth screen shot (Image courtesy of Liz Earley. Map data: Google, Get Mapping plc)

Alternatively, the students use digital cameras to produce short documentaries or edit their own video podcasts using *Splice* at the site location. These videos required the students to explain the world around them, thinking carefully about the landscape they see; giving lots of opportunities for Assessment for Learning (formative assessment). These videos are structured around specific learning tasks, aiming to enhance their usefulness as a learning tool and improve the narrative of the video output.

Technology required:
Hardware required: iPads, Griffin Survivor iPad case, digital cameras (preferably with GPS capabilities)

Software required: *Skitch* (also available on Android), *GPS Log,, Fotobabble, Polldaddy*.

How much time was taken to develop this and implement the method?
Approximately 4 days of time spread over a year, spent developing these activities, knowledge of the technology and how to integrate this approaches into current fieldwork investigations.

What evidence is there of the effectiveness of this activity?
These methods are presently used by both Geography and Biology courses to enable students to create topic-specific videos focusing on landform formation and data collection techniques. This stimulates focused-thinking and evaluation in the field and provides useful resources for student reflection after the fieldwork data collection has ended.

Feedback from visiting staff with groups:

iPads in the field used in innovative & engaging ways
The use of technology certainly enhanced the fieldwork experience and quality of work produced

Staff at the field centre are currently looking into how we can use these apps and videos to support learning further- critically evaluating what the current learning outcomes are and how student attainment can be improved.

Pitfalls/Problems/Limitations:
Initial investment in iPads is high and the devices can accrue some maintenance costs when frequently used in an outdoor setting. Lack of connectivity can be an issue-particularly when trying to get a large amount of data on/off the iPads in a short amount of time. Plugging iPads into laptops to transfer photos & videos rather than using storage systems such as *Dropbox* has made this process quicker. However it is not possible for all file types e.g. kmz files from *GPS Log* have to be emailed.

Technological expertise required:
Staff skills required: Staff need to be confident using the iPads and the specific apps. Some online form development is needed prior to using *Polldaddy*.

Student skills required: Most students are up to speed when it comes to these technologies and it is important to remain in step with them. It is great that we can show them alternative uses to support their understanding of the outdoors using platforms that they are familiar with.

Key tips or advice to others:
In order to use iPads effectively and for all students to be fully engaged ideally students should have an iPad per 3 or 4 students so a significant investment is needed. Focusing on the learning objectives for the session and considering how the technology can be used to meet these, rather building activities around apps allows the devices to be useful learning aids rather than just gimmicks.

Don't try and do too much! While brilliant at engaging students and encouraging them to look critically at their surroundings there is a danger of students spending too much time looking at a screen rather than the world around them.

Also see: Wills and Earley (2013)

The next case study is a simple case of students doing some research on a fieldwork area before going into the field and learning some basic presentation and information literacy skills. It does not require a tablet but students could produce a

3.2 Drawing and Painting Tools

poster reporting their fieldwork after the event with photographs taken on an iPad. iPads have all the necessary apps (Pages/Keynote) and camera plus photo library, to make a poster on a tablet and project it to a group for example.

Case Study 4: Student-Produced Posters of the Fieldwork Area (Brian Whalley)

Institution: Queen's University of Belfast, UK

Keywords: Electronic poster, preflight exercise

Aim: To provide an introduction to a fieldwork area by getting students to research the topic and provide a poster(s) about the area.

On this module the students had to do posters prior to going into the field. This gave them an exercise in researching from various resources and selecting them according to a brief for the poster location (a field study centre). The task was assessed using published mark scheme criteria. The case study task was for students to research about the fieldwork area. For this area they needed to research;

1. The geology of the area specified
2. The geomorphology of the area;

A poster was necessary for each.

This practical was worth 10 % of the module. It was their first activity at this Level (second-year) and they have to learn new skills as well as research techniques. This practical was set in the first week of a semester; thus, they get marks early in the module and have a sense of achievement and learned something about the field area beforehand.

As well as students sending in their posters they also (if they wished) entered their posters in a competition. On the first evening of the field trip students handed in A4 copies of their posters from which their name has been removed. These are then stuck on the walls of the laboratory and numbered. At a suitable time the next day students voted for their 'best' poster (criteria can be discussed in advance) and the one with the most points wins. This sets me back a bottle of wine as the prize. This is one way in which students see what others have done and how they can be improved, stuck to the guidelines etc. One advantage of this is that you do not have to give students lectures on the topic; they do the research themselves; staff only have to spend time marking.

Technology required:
Students produce A4 size posters of a given topic or remit using *PowerPoint* or similar application such as *Keynote*. The module used individual production but a poster could be produced as a collaborative effort, on the field trip itself and downloaded to a tablet as appropriate.

Information about how to do it, expectations and criteria were posted on the module website and examples of various posters were available for inspection on the module website.

Approximate cost of technology:
A few pence, even if colour printout of an A4 page is used. The assessed posters are e-mailed in. Production of a PDF from PowerPoint or Pages gives an extra skill.

How time intensive is this to set up and implement?
This activity was done in advance of the fieldwork time by students. They did the exercise as an assessed piece of work. In effect, it acts as a 'preflight' (see Whalley 2013) for the weekend activities so less instruction time is required.

What evidence is there of the effectiveness of this activity?
As part of the module, students are asked to reflect on their practical work as part of a portfolio. Student comments as the two below:

> "I found this a useful exercise as it allowed me to get a really good idea of the area I was going to be seeing in the field trip and therefore allowed a greater depth of understanding while in the field. I realised when all the posters were displayed that I hadn't entirely done what was asked and that I should read the requirements before starting and after finishing a piece of work."
> "An enjoyable practical which allowed me to gain and further develop my skills with regard to *PowerPoint*, image editing and researching information on a relatively small area—which was challenging. I was quite pleased with the work that I produced from it and I found the preflight beneficial here as it gave me a starting point to develop from. I was pleased with my mark and the feedback and deemed it reasonable as I thought myself that the information I had presented in the poster could have been further developed."

Pitfalls/Problems/Limitations:
The only problem is making sure that students can make posters in the chosen app (*PowerPoint*, *Pages* or *Keynote*) and that they have some basic design skills.

Technological expertise required:
Staff skills required: None, only in providing criteria for the assessment.
Student skills required: Researching the topic, making judgements about source validity, what to include and what not, poster and graphic design, use of software.

Also see: Whalley and Rea (1998) and Whalley (2013)

The next case study is a rather more complex version of student viewing materials for a specific area of investigation. Gary Priestnall's project involves fieldwork but based around analysis and applications using 'desktop' computers. Note that, at present, *ArcGIS Online* processing seems best on Windows OS or cloud-based processing. It is currently rather restricted on iOS. Nevertheless, technologies will continue to converge allowing true field processing in the future. Until then, this case study shows ways of landscape modelling. *Bryce* is a 3-d modelling package (available for Windows and Mac OS, Bryce version 7 is now available), alternatives exist and 3-d terrain generators (such as *Tarragen*) could be usefully employed.

Case Study 5: Landscape Visualisation in Fieldwork (Gary Priestnall)

Institution: University of Nottingham, UK

Keywords: Landscape visualisation, fieldwork, Digital Surface Models, representational fidelity, geographical information systems.

Aim: To use 3D visualisation models to better understand and interpret the glacial landscape

Visualisations of the glacial ice model were created using the *Bryce 5* modelling and visualisation package. Having used visualisation to support the discussion of the landscape's history and development, a project was designed to engage students with the issues relating to the digital representation of landscape.

This teaching method involves a field course exercise designed for first year geography undergraduates. It takes place over a four day residential course and is a core geography module covering both physical and human geography. The exercise's aim was to specifically engage students in the comparison of landscape visualisation and the real scenes. Acquiring spatial knowledge through fieldwork is considered to be very important for integrating separate observations onto a common frame of reference. The students had to use their materials to find the locations of the static views (photographs), and compare them to the reality of the landscape.

A series of perspective views of the terrain draped with geology was used in orientation exercises to describe the broad character and history of the landscape, including the glacial history of the region. Visualisations of the glacial ice model were created using the *Bryce 5* modelling and visualisation package. These visualisations were static images which were printed out to support the field-based orientation exercises. Having used visualisation to support the discussion of the landscape's history and development, a project was designed to engage students with the issues relating to the digital representation of landscape.

This project was designed to take a full day including the written report. The aim was to enable students to produce reasonably believable landscape visualisations using commercially available data sources and then to compare the outputs with their real-world counterparts.

The learning objectives of the project were:

- To assess the degree to which individual computer-generated scenes derived from radar terrain data and aerial photography is representative of the real scenes.
- To develop a schema for evaluating such landscape visualisation by outlining the factors that proved important when making the comparisons.
- To assess the degree to which evidence observed in the field supports the computer generated 3D reconstructions of glacial ice retreat.

The project started with a briefing where students were organised into groups of four. The concept of a grid based Digital Surface Model derived from airborne radar was introduced, and it was explained that landscape visualisation is often used to represent real scenes in public contexts.

The students were then asked to select five appropriate locations which they would use to assess the strengths and weaknesses of the techniques and data for representing real scenes. The students then used these location viewpoints to create their 3D views using the Bryce5 package. The view was printed out and the glacial model was merged into the scene and printed onto acetate to enable a simple 'augmented reality' technique to be used.

The next step was to visit each of the viewpoints in turn assisted by GPS if necessary. The computer generated printouts were annotated and field sketches were made, this was done to evaluate each of the images' sense of reality. Photos were also taken so the real life picture could be compared with the computer generated image.

The printouts of the glacial scene were used to convey the position and extent of the possible scenario for glacial retreat. Back at the field centre the groups had to develop a structured schema by which the computer generated visualisations could be assessed against the real scenes they observed, and to illustrate this with evidence gathered in the field.

Technology required:

Digital models and 3D visualisations of the Earth and its surface are becoming increasingly common in the public domain. There are also increasing expectations of what is possible, these expectations are driven by the computer games industry as well as companies such as Google and Microsoft, whose virtual globes (*Google Earth* and *Virtual Earth 3D* respectively) are becoming increasingly popular. They are not only key tools for geographers, as they are widely available to the public for their use as well.

There are various methods and technology which can be used to create digital models. Data sources for elevation include direct ground survey, contours, spot heights, photogrammetry, airborne laser measurement known as Light Detection And Ranging (LiDAR) and Interferometric Synthetic Aperture Radar.

There are many methods of producing the materials used for the exercise, using various types of technology. But the technology used in this exercise was Interferometric Synthetic Aperture Radar; this was used to create the digital surface model. Aerial photography and other forms of visual aids were also obtained.

The other technology used included *Arcview* GIS and the *Bryce 5* modelling and visualisation package. These were used to create the surface model of the scenario with glacial ice and moraine deposits. GPS was also used to navigate in the field, and cameras are needed to take pictures in the field.

How much time was taken to develop this and implement the method?
The whole course took place over four days, but the main fieldwork project took a whole day to produce both the imagery as well as the written report afterwards.

What evidence is there of the effectiveness of this activity?
This teaching method has been found to be very successful in engaging student groups in digital surface modelling and visualisation at an early stage in the undergraduate curriculum.

As landscape visualisation is being used more and more in public contexts, the emphasis from a teaching and learning perspective has been encouraging a critical awareness of where these visualisation techniques could/should be used to represent reality in a true manner.

Pitfalls/Problems/Limitations:
Some specific observations emerging from student reports, feedback forms and group discussions over a number of years have highlighted a number of problems. These include scale issues where the computer generated image is poorly presented compared to the real world picture.

Technological expertise required:
Staff skills required: Must be able to use the technology used. They must have a good understanding of various technologies including Interferometric Synthetic Aperture Radar, the Bryce5 modelling and visualisation package and knowledge of GPS. They will need to be capable of using cameras.

Student skills required: Students must have a good understanding of all the technology that they will be using; the staff will have to assist the students with the creation of the computer visualisations. Students will also need to be capable of taking photographs in the field. NB The latest version of Bryce is now Bryce7 but this is not yet available for Mac OSX beyond 10.6 nor is there an iOS version.

Also see: Priestnall (2009).

The previous case study mentioned 'assessment'. This is a major aspect of education from school to university. How do we reward practise with marks, especially for fieldwork where an examination is hardly appropriate? The next case study helps to show how alignment of assessment is important. In this example, note that 'mediascape' can be used in a general or generic sense. It can be a digital media artefact or combined artefacts associated in some way with spatial areas or regions. They can be triggered in the mind by the location of the person experiencing the media. Thus in a mediascape a person may walk around an area and as they do so they will hear digitally stored sounds associated with different places in that area. This is a linked concept to Augmented Reality (AR) mentioned below.

Case Study 6: Technology-Mediated Field Learning (Claire Jarvis, Jennifer Dickie and Gavin Brown)

Institution: University of Leicester, UK

Keywords: Mediascape, teaching, learning, mobile technology, field course

Aim: For students to use and develop mediascapes of Dublin field area

During the field course mediascapes were included within the traditional field teaching methods, these included staff-led tours, discussions and student-led group research projects.

This teaching method took place during a 2nd year Human Geography field course to Dublin. The teaching method includes mediascapes using the freely available software *mscape*. (See note the end of the case study.) During the field course mediascapes were included within the traditional field teaching methods, these included staff-led tours, discussions and student-led group research projects.

The mediascapes performed two roles within the field course. Their first role was to act as a guide for the students; this was in addition to the staff-led introductory tour. Secondly, the students had to develop and produce their own mediascapes as an expression and reflection of their findings in Dublin. To do this the students had to collect photographic and sound-based material in small groups.

The field course was assessed using a group presentation. The students were also assessed individually, as they had to produce a project report consisting of the group mediascape, as well as individual reflective essays.

Technology required:
Mobile technology assisted teaching and learning is relatively new within Geography and other subjects, but its use is rapidly increasing. Mobile technologies offer an efficient replacement to traditional teaching and learning methods, it also provides students with the opportunity to practise their basic generic skills.

This teaching method incorporates mobile technology to produce mediascapes. The mediascape is composed of sounds, images and video footage. A GPS-enabled handheld computer (PDA) is used to produce the mediascapes. The freely available software mscape was also used.

Approximate cost of technology:
A GPS-enabled handheld computer is used; an example of this is a Personal Digital Assistant (PDA). A single PDA generally ranges from £100 to £2000 depending on the type and quality.

What evidence is there of the effectiveness of this activity?
The student's opinions of the mobile technology during the field course were that overall, the mscape assignment allowed a more complete expression of their ideas than text alone. The general consensus from the students was that the mscape assignment was much better and much more enjoyable than just writing an essay.

Pitfalls/Problems/Limitations:
The main difference between this method and traditional essays is that the mscape assignment lacks structure. Studies have shown that students who adopt a deep learning approach form highly structured knowledge. However the mscape method does not foster this. However, the mscape method allows free thinking and deeper learning.

Technological expertise required:
Staff skills required: Must be able to use the PDAs and instruct students how to create mediascapes using the mobile technology.

Student skills required: Need to listen to staff and their instructions on how to use the PDAs and any other technology required.

Also see: Jarvis et al. (2010)

Note: Updated software is now available to achieve the construction of mediascapes using *AppFurnace* (2014) for publication on iPhone or Android platforms.

3.3 Summary

This chapter provides a number of case study examples on how to bring visual imagery into field teaching in variety of active ways from the very simple and the use of the *Skitch* app in Case Study 3 to the more complex use of 3D visualisations in Case Study 5. It is hoped the range of visual imagery case studies will inspire confidence for tutors to adopt some of these visual elements into field teaching.

References

AppFurnace (2014) Creating iPhone and Android apps. Retrieved from http://appfurnace.com/
E-fieldwork (2014) Enhancing fieldwork learning Pinterest website. Retrieved from http://www.pinterest.com/efieldworkl/
GEES-projects (2006). Retrieved from http://78.158.56.101/archive/gees/projtheme/smallfund/2006/projs06.htm#evalrichmed
Goodliffe A (2011) How the iPad transformed my teaching. Retrieved from http://www.as.ua.edu/ipad/dr-andrew-goodliffe-how-the-ipad-transformed-my-teaching/
Jarvis C, Dickie J, Brown G (2010) Aligned assessment of technology-mediated field learning. Planet 23:68–71. Retrieved from http://78.158.56.101/archive/gees/planet/p23/P23.pdf
Priestnall G (2009) Landscape visualisation in fieldwork. J Geogr High Educ 33(1) 104–112. Retrieved 7 Dec 2011 from http://www.tandfonline.com/doi/pdf/10.1080/03098260903034020
Whalley WB, Rea BR (1998) Two examples of the use of 'electronic posters'. J Geogr High Educ 22(3):413–417
Whalley WB (2013) Teaching with assessment, feedback and feed-forward: using 'preflights' to assist student achievement. In: Bilham T (ed) For the love of learning: innovations from outstanding university teachers. Palgrave, London, pp 97–102
Wills M, Earley L (2013) Slapton Ley and Rhyd-y-creuau: outdoor technology, vol 43. FSC Magazine, Spring 2013, pp 12–13

Chapter 4
Display and Recording: e-Books and Field Notebooks

Abstract In this chapter we look at ways of providing teaching materials for student field use. We consider the creation of a digital field (or laboratory) notebook and include experiences of using a variety of apps for taking notes in the field. We also devote a separate section to PDF file handling.

Keywords Digital notebook · e-book · Geo-referencing · Field notebook · PDF

4.1 Fieldwork Materials

Previously, materials for fieldwork might well be instructional books several pages or even tens of pages long. These were printed off, at time and expense, well in advance for distribution before fieldwork. This can now be done by editing and selecting from electronic media and distributed electronically by e-mail or via *Dropbox* or PDF handling apps such as *PDF Expert*. Although such documents are probably constructed in 'word processing' packages such as *Microsoft Office*, it is probably better to produce PDFs from these to make printable documentation. Operating systems now make PDF production easy. Functionally, it is worth noting that PDFs intended for electronic distribution, especially by e-mail, should be reduced in size (compressed) before sending. As documents for fieldwork may well include diagrams and photographs, file sizes can be considerable even before PDF production. As well as *Adobe Acrobat Pro*, other facilities for file size reduction are available both on and off-line. As an illustration, a 4.6 MB coloured conference programme was reduced to 93 KB! When students produce their own reports, for example, they should be made aware of file size bloating and the means to reduce it. File size reduction also applies to presentation apps such as *Keynote* and *PowerPoint* when posters are produced.

4.2 e-Books and Related Formats

Recently, various formats of 'e-books' have become available as downloadable material to tablets. These are in general use via *Kindle*, *Zinio* (for magazines) and bespoke apps such as *The Guardian* newspaper. As well as the educational aspects in such books, *iTunes U* provides specific educational resources that might be useful in fieldwork. For example, the University of Sheffield Archaeology Department has iDig—Field training for archaeologists (http://www.idigsheffield.org.uk/). Several universities now use *iTunes U* for displaying their research, publicity and educational materials. There is no reason why students' work, especially student projects, could not be presented in *iTunes U*. However, preliminary results suggest that students need to know about intellectual property rights because *iTunes U* is a global platform under the auspices of a specified institution.

Tutors may want to use the facilities provided by free apps to produce their own e-books. By this we mean integrated compendia of materials that not only can be read on tablets but which provide table of contents, hyperlinked pages, text size changes as well as readability. A number of apps are available for the creation of e-books or e-pubs. Some are desktop computer based (such as *iAuthor*) others, such as *DemiBooks Composer* can be used on a tablet. *Calibre* is, at the moment, a popular desktop application for making e-books (EPUB and PDF formats) as well as making conversions for publishing to Kindles but not (yet) for making the documents on tablets as with *Composer*. However, such apps are increasingly easy to use and there is no reason why students cannot be set tasks producing and using their own e-books, perhaps as interactive fieldwork reports or even dissertation projects. This would give them useful digital and information literacy skills.

We think electronic publishing is an area ripe for development by both students and tutors. For example, the special interest group on Media-Enhanced Learning (MELSIG) produced 'Digital Voices' edited by Andrew Middleton (see Middleton 2013) and can be found in printed and e-book versions, that also gives a *Slideshare* presentation as an alternative form of information distribution.

4.3 PDF Manipulation

There is a surprisingly large number of apps dealing with PDFs and which offer a variety of facilities. We list the most popular: *GoodReader*, *Readdle*, *PDF Expert*, *PDFReader*, *Annotate+*, *Readability*. These apps can act as filing systems for reading (such as *GoodReader*) but their main advantage is that as well as providing a common source for several apps (such a word processors and presentation tools) they can be searched and annotated. Most apps provide help documentation or on-board manuals. Although there is not space to discuss further, tutors should be aware of the possibilities of using tablets to link PDFs to references via bibliographic tools such as *Endnote, Zotero and Mendeley*.

4.4 Field Notebooks

We now continue by looking at ways in which fieldwork notes or data can be collected using iPads. We have already shown how some drawing tools, such as *Skitch*, can be used to add information to images. In fact iPads are ideally suited for this role, as is shown by the many notebook apps available within the App Store. It is not possible to list them all, let alone review them, so we are just indicating the types that are available, noting that some are extensions of existing applications for 'desktop' or 'laptop' computers. Several notebooks are available in free basic ('lite') modes with the possibility of upgrading to 'premium' versions. The latter might have fees per month or per year or storage limits. These need to be explored for your own project suitability. *YouTube* may also have useful, 'how to' videos.

iPads, like other tablets, comes with a note-making app, *Notes*. Although this is rather basic it can be called and the contents searched via iOS from the home page. The pages are date and time stamped and can be set up so that notes made in the field can be synced with an OSX computer on your home Wi-Fi network. It may be all you need to do is cut and paste material into another app, such as *Pages* or *Keynote*.

Notability is a typical and commonly used note-taker which allows typing and sketching as well as voice note taking on an integrated page. It can also add 'sticky' notes and images to your page. It also supports hand writing (although you would not want to do too much of this) and a wrist guard area makes sure you don't smudge the page. Notes can be opened in some other apps and converted to PDFs and mailed or exported.

Evernote is a longstanding application that is essentially a scrapbook that allows accumulation of text, images, web pages, voice memos, e-mail clipper etc. and organise them. This is a very capable cross platform application. *Skitch* comes from the same stable as does *Penultimate*, which is designed for pen-on-paper note taking, drawing and integration.

Some other notebook apps are *NoteSuite*, *Inkpad*, *Moleskine*, *Irisnote* and *Paper53*. *Memo* is an app that can be used as pen/stylus note taker but which can also export what you have written as an OCR (optical character recognition/handwriting recognition) file. Several mind-mapping apps (and similar) are available. Examples are *Total Recall*, *MindMeister* and *SimpleMind+*. Under some situations, pure audio note-taking might be the best way. There are several apps available e.g. *Recorder Pro* as well as the note-taking apps mentioned previously.

4.4.1 Livescribe Smartpen

The Livescribe Smartpen has been on the market for several years and has been used for fieldwork where students need to take extensive hand-written notes or make sketches. It uses special 'dot' paper (available in various sizes of notebook) and a ballpoint pen to make digital notes. These written notes can, if required, be accompanied by audio recording. The written and audio materials can be uploaded onto a desktop/laptop via a USB cable link. The latest version, Livescribe 3 can mate with Livescribe+ to sync with a smartphone or iPad. Although this is expensive (£150), it might be a very useful device for making audio-enhanced field notes or for students with certain accessibility difficulties.

We now turn to data collection in the field using dedicated applications. Alice Mauchline and Rob Jackson used a number of apps for their microbiology fieldwork in Iceland and report on their students' experiences.

Case Study 7: iPads as Digital Field Notebooks in Microbiology Fieldwork (Alice Mauchline and Rob Jackson)

Institution: University of Reading

Keywords: biosciences, iPads, *GeoSpike*, *GPS Log*, *Splice*, *iMovie*, *Yammer*, *Google Drive*, *Dropbox*, *Facebook*, *Storify*

Aim: To use mobile technologies to support student learning pre, peri and post fieldwork

The University of Reading runs a microbiology fieldtrip, led by Rob Jackson, to Akureyri, Iceland for students to gain skills in microbiology research; in particular to study microbes that inhabit extreme environments such as glacial rivers and areas of geothermal activity. The students spend several days sampling the environment and then spend 10 days culturing and characterising the microbes in the lab.

In 2013, the fieldwork research activities were supported through the use of mobile technology and social media to engage students in their learning and to facilitate communication between the students and staff in the large multi-national team.

Mobile technology and social media was used to support learning in different ways in the pre, peri and post-fieldwork phases. An enterprise social networking site (*Yammer*) was used to facilitate communication within a closed network and for collation of literature and lecture materials; it was also used for disseminating preparatory material prior to the trip. While in the field, the students conducted their sampling in pairs and each pair was provided with a ruggedised iPad loaded with many fieldwork-related apps. The students were directed towards several apps to capture field data while sampling in extreme environments.

Students were encouraged to collect geo-referenced, real-time data at each sample position (using *GPSLog*), which included photographs, videos and field notes of environmental conditions. These data were captured offline while in the field and stored. Once back from the field and connected to the internet, the students shared their field data using cloud storage (*Dropbox*) and a 'live' database of results (using *Google Drive*) to link to their laboratory experiments.

4.4 Field Notebooks

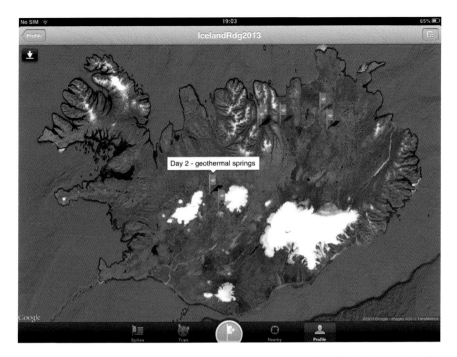

Fig. 4.1 Screengrab of sample locations recorded using *GPS log* then plotted and shared using *Geospike*. (Map data: Google, TerraMetrics)

The geo-referenced sampling positions were shared using a social travel log website/app (*Geospike*) for long-term data retrieval (Fig. 4.1). The photo and videos were used in their assessed presentations and to make short reflective videos of the sampling environments (using *Splice*, *iMovie* and others). Social media was used extensively for informal communication (*Facebook*) and reflection (*Twitter*); see this *Storify* for a summary tinyurl. com/ovksh9f. Many of the students further explored the use of the iPads and apps for their learning and found them useful in ways they were not expecting e.g. to display laboratory protocols, for recording lecture notes.

Technology required:

- A set of iPads with Griffin Survivor cases
- A selection of apps

What evidence is there of the effectiveness of this activity?
There were some drawbacks; including concerns over breaking the equipment (which didn't happen!), finding it cumbersome at times and difficulties sharing them out. However, overall, the feedback was very positive from the students. Some examples of the benefits they identified were;

- 'the iPad was exceptionally useful for the fieldwork; recording data and geographical info together'
- 'brings together several useful applications in one place e.g. GPS, photo/video/internet, so we did not need 3–4 pieces of equipment'
- 'instant note-taking, no faffing about with pens and paper. All pictures, recordings etc. in one place, can use and manipulate data instantly'.

In conclusion, mobile technology provided a more efficient and accurate way to capture, store, analyse and share data during fieldwork. Further, it provided a means to establish good working relationships between the students and helped team cohesion between both nationalities.

Technological expertise required:
Staff skills required: Should familiarise themselves with the apps before the teaching session, but most staff members with a basic understanding of iPads should be able to use this technology.

Student skills required: Before beginning the fieldtrip make sure that all students are comfortable with iPads.

Also see: Jackson (2012) and France et al. (2013).

4.5 Geo-referencing and Geo-tagging

Geo-referencing, or geotagging, is the allocation of geospatial metadata to digital media, i.e. images, videos, audio or a data file. The trend has infiltrated social media sites such as *Facebook*, *Twitter* and *Flickr*, each using geo-referenced photographs and status updates and has generated a wave of 'geo-awareness' (Luo et al. 2011). The technology used to geo-reference can include anything from a digital camera with a Bluetooth-connected standalone GPS receiver to an iPad to a complex GPS station.

Geo-referenced metadata can be manually connected to a file, i.e. in apps such as *Flickr* or *FieldtripGB*, or metadata can be automatically applied using a specific hardware, such as smartphones. When manually geo-referencing, it is essential to attach meaningful labels to the media or data file as these labels describe the media for the next user and for future reflection. As the geo-reference only refers to the coordinates of the media, labels should include the viewing direction of photographs, travel trajectory and subject-specific information that you need about the time of day, etc. Where GPS capabilities are available on a smartphone many social network sites, including Facebook, can utilise this facility automatically; otherwise users can revert to manually referencing their location (Luo et al. 2011). This technology opens up an opportunity for in-the-field data collection and quasi-real-time follow up investigation on site (Traxler and Wishart 2011). "The integrated use of geo-referenced data is invaluable for studies on mobility, and the unique ability of mobile devices to stay 'inworld' with the participants is the key attraction of using this technology for research" (Beddall-Hill et al. 2011).

4.5.1 Problems and Limitations with Geo-referencing

Geo-referencing, particularly asking students to tag their locations on social network sites, can have related privacy issues (Friedland and Sommer 2010). A small minority of students who attended field courses with the University of Chester, Department of Geography and Development Studies in 2012/13 were reluctant to use personalised geo-referencing apps, because the general public may then be aware that the students were away from home. This perceived risk was not held by most students, but the University accommodates for the minority by providing Department log-ins; creating a perceived anonymity at the same time as additional control and supervision of public feeds by staff.

Geotagging can have 'noise' related issues, as GPS sensors or manual tagging can have attached inaccuracies. For example, a quick trial using a geo-referencing app on an iPhone and a Samsung Galaxy Mini found that the iPhone came with more reliable GPS capabilities with the Android being inaccurate by over 10 m. These devices were trialling the same app, but there were several possible reasons why the GPS accuracies varied—operating system, app capabilities, and hardware GPS capabilities. This test, although subjective and uncritical, illustrates that smartphones can vary significantly in their accuracy.

Our next case study, from Katharine Welsh, shows how a simple app, *Flickr*, which is commonly used for image collection, can be used effectively for 'geotagging'.

Case Study 8: Geotagging Photographs Using *Flickr* (Katharine Welsh)

> **Institution:** University of Chester
>
> **Keywords:** geotag, smartphone, photographs, spatial awareness
>
> **Aim:** To capture and share geo-referenced fieldwork photographs through the *Flickr* app
>
> GPS capabilities on smartphones, tablets and digital cameras now make it very easy to tag photographs to digital maps and use these resources to support accurate data collection. The technology is simple, free and can be used in a range of fieldwork settings.
>
> Five Level Four students (first year undergraduates) worked on a small group project looking at changing house prices in relation to the shoreline from Slapton Sands to Kingsbridge, South Hams, Devon. As part of their data collection the students used geotagging to add a spatial reference to the photographs that they took, as part of their evidence base.

Rather than using specialist equipment, one of the students downloaded the *Flickr* application from the iTunes App Store on his iPhone4. The students were able to use the *Flickr* app to take photographs of the houses, add a title, geotag and a description onto the photograph. Back at the field centre, the students created a free account for the desktop/laptop version of *Flickr* and the geotagged images were plotted on a map for them.

The learning outcomes were a greater awareness of geotagging visualisation of spatial data, basic understanding of GPS and an awareness of spatial accuracy of smartphones. The overall assessment from this activity was to produce a digital podcast (i.e. focusing on either their methods or results) and integrate that into their written report.

Technology required:
Flickr is currently available on Android, Windows phone, m.Flickr.com, iPhone or iPad. Other geotagging software apps such as *GPSLog*, *Panoramio* and *Evernote* may be suitable for the same purpose.

Approximate cost of technology per unit:
Cost of phone will vary and app is free.

How time intensive is this to set up and implement?

It took approximately five minutes to download the app from the iTunes App Store and a further five minutes to set up a free *Flickr* account on the desktop/laptop version. Staff involvement was minimal and the data collection was student-led despite the students having no previous knowledge of how the *Flickr* app worked.

What evidence is there of the effectiveness of this activity?
Staff members who were involved with this project were impressed by how the students led the way for using the technology.

The students had a greater awareness of device accuracy—there were problems with GPS availability—and they felt that they could take this skill to an employer and apply it confidently with more sophisticated software/hardware. The students now have a much better understanding of how mapping data spatially can be useful for visualisation of patterns, which was not previously understood. It was also useful to show them how their personal hardware (iPhone) could be used successfully within a learning environment and that many useful applications that could aid their learning were free.

All of the students who used geotagging in this way felt sure that they would use it again in a future project or use a similar technique with their future potential employers. Although for this study, geotagging photographs was not used as a method of post-fieldwork reflection, it certainly could be used in that way as an accompaniment to reflective field diaries which help to facilitate deep learning.

Pitfalls/Problems/Limitations:
The data were largely geotagged accurately; however, there was one anomaly. This anomaly was used in a productive way to show the students the potential inaccuracies associated with this method. While the data were accurate to several metres, which was fine for this project, this may not be suitable for those who required more accurate results for small-scale spatial changes. The students felt that it would have been useful to save the data points and transfer them to *Google Maps* for further analysis, but at the time the students collected the data the only option was to view the data in the *Flickr* map. *Panoramio*, *GPSLog* and other geotagging photographs apps allow the data points to be used in *Google Maps*.

Technological expertise required:
Staff skills required—Could be negligible if students are able to use their own devices.

Student skills required—ability to use (and accessibility to) a smartphone.

Also see: Welsh et al. (2012) and Welsh and France (2012).

The next three case studies use dedicated apps to collect field data. These apps are primarily for multiple workers all collecting the same kind of data. As with this type of scientific endeavour, care must be taken well in advance for the success of the project.

Case Study 9: Using Mobile Applications to Collect Field Data: Application of EpiCollect (Charles Harrison, Paul Wright and Rhu Nash)

Institution: Southampton Solent University

Keywords: Smartphone, mobile applications, geotagging

Aim: Using a citizen science project and *Epicollect* to establish a database of ancient yew trees

This project uses a free, open source platform, called *EpiCollect* (2014). Through the website a project can be constructed, which can be accessed using the free *EpiCollect* Smartphone Android or iOS app.

In our case, this application was used to establish whether it was an appropriate vehicle for developing a citizen science project to establish a database of ancient yew trees; however, the application can be used for a myriad of field studies, provided that the accuracy of the on-board GPS is not seriously tested. The use of the mobile app allows the project team to use a range of social networking tools to establish a community of 'measurers', who can all report data back to the team, affording data collection on a spatial scale not always possible in conventional fieldwork.

The *EpiCollect* user/project team has two options:

First, establish a database of geo-referenced photographs, using the phone's GPS to position the sample. Or secondly, make the data collection more sophisticated by developing a data response form, which is uploaded alongside the photograph. This is easily done using the 'drag and drop' set up page on the *EpiCollect* website. Options available at present are short and long text input, and the selection of single or multiple responses (such as Likert scales).

Data can be collected without the need for a mobile network, as a database is developed and stored on the device until the device returns to a Wi-Fi network where the data can be uploaded. The user can access their complete dataset through the unique project URL given by *EpiCollect*. These data can be saved in XML or CSV for further analysis, viewed in its tabular format, as a graph or the user can use the *Google Maps/Google Earth* mapping interface to view the data. The map can be manipulated to show a timeline of data points, and each point can be selected to reveal its associated photograph and other data fields.

What technology is required?
A list of compatible smartphones is at *Epicollect* (2014). Access to a computer/tablet and the Internet is required to set up the project and view data.

With the growth in the smartphone market predicted to rise, and the predominance of iOS and Android platforms within this market, it is envisaged that this technology will become more prevalent within the student community.

Approximate cost of technology:
The key here is to use the technology that many students already have to hand, rather than it being something that should be provided.

How much time was taken to develop this and implement the method?
The setting up of a project form takes less than an hour. The data collection takes place as usual in the field. Data are synced when access to a network is present, and is viewable almost immediately.

What evidence is there of the effectiveness of this activity?
The uptake of the app by project groups was small and slow for two reasons. First, the time it took to spread app awareness and adoption was greater than the period of study. Secondly, the volunteer groups who did use the app had a demographic that was much older than the demographic of smartphone use. However, within a student fieldwork context, these concerns may not apply.

Reflections on app use showed that *EpiCollect* was easy to set up, and the application was praised for its intuitive design. Most users felt the app easy to manipulate, although some students commented on the fiddly nature of their smartphone's keypad. Again, if students are using their own phones, this familiarity should enable quicker and smoother usage.

Pitfalls/Problems/Limitations:
Students may not have the technology. This is particularly true of the other platform users, such as Blackberry or Symbian (Nokia).

GPS accuracy was checked against a handheld Garmin GPS. There seemed to be good agreement. However, mapping phenomena over relatively small spatial scales, e.g. a few metres apart, may not be possible.

There is a potential issue with wet weather and battery life.

Technological expertise required:
Anyone conversant with the internet, and using a smartphone, should be able to make use of this technology.

Also see: Epicollect (2014), Aanensen et al. (2009) and Harrison (2011).

Case Study 10: EpiCollect: Linking Smartphones to Web Applications for Epidemiology, Ecology and Community Data Collection (David Aanensen[1], Derek Huntley[1], Edward Feil[2], Fada'a al-Own[2], Brian Spratt[1])

Institution: [1]Imperial College London and [2]University of Bath

Keywords: *EpiCollect*, Biology, Epidemiology, Ecology, data collection and analysis, database, mobile phones, Smartphones, Android, fieldwork, web applications, GPS, *Google Maps/Earth*, mapping, visualisation.

Aim: To enable epidemiologists and ecologists to collect field data through their mobile phones

Recent advancements in mobile technology and the introduction of Android phones have created new opportunities for developing mobile phone applications. By combining these new mobile phone applications with web applications, it can allow two-way communication between field workers and their project databases.

Epidemiologists and ecologists often collect data in the field and, on returning to their laboratory, enter their data into a database for further analysis. The recent introduction of mobile phones that utilise the open source Android operating system, and which include (amongst other features) both GPS and *Google Maps*, provide new opportunities for developing mobile phone applications which, in conjunction with web applications, allow two-way communication between fieldworkers and their project databases. Data variables can be entered into *EpiCollect*, for example, soil pH, temperature and moisture. Data and photographs can be submitted via *EpiCollect* from the phone to the project database.

EpiCollect has potential for the collection and submission of data from a single fieldworker, as well as the phone's ability to view the data retrieved from the central database using *Google Maps*. However, *EpiCollect* was designed for situations allowing multiple field workers to collect and submit their data from various locations. The software works particularly well when used by multiple field workers, and it is especially useful with the younger generations as the mobile phone is generally their natural way of transferring information.

Technology required:
The technology used includes mobile phones that utilise the Android operating system, with GPS and *Google Maps/Earth* as well as other features and the teaching method revolves around the software *EpiCollect*.

Approximate cost of technology:

- *EpiCollect* software—free open-source project

How much time was taken to develop this and implement the method?
Users are able to utilise the free software *EpiCollect* to create their own projects; therefore, the time required to develop the project/s varies.

What evidence is there of the effectiveness of this activity?
Although *EpiCollect* was originally designed to allow the collection of epidemiological data, it also has the potential to be used in other situations. For example, it could be used in biodiversity research, where multiple researchers could record details and photograph the presence and distribution of a particular species. This approach could be used by the Zoological Society of London's EDGE of existence programme, which is the only global initiative to focus specifically on threatened species (see Edge 2014).

The team that created *EpiCollect* have also highlighted the software's other benefits, such as the *EpiCollect* mobile application does not require mobile network access for data collection. Users can collect data in remote areas using their phone's GPS and camera, which they can use later to synchronise the data when they return to an area that has mobile network coverage or when connected to a wireless network.

Another advantage of the software is that there is no limit on the amount of projects that can be set up by users, as long as the projects are not deemed as abusive, offensive or contain any illegal data. The team is also looking to develop the software further allowing users to create more complex projects.

Pitfalls/Problems/Limitations:
EpiCollect's major potential use is for field surveys in resource-limited settings that collect key information to inform important decisions regarding resource allocation. But there are a number of current issues centred on cost and mobile data network availability that could limit the use of systems such

as *EpiCollect* in these settings (e.g. parts of sub-Saharan Africa). Although the data can be stored on the phone's database and synchronised when the user returns to an area with data connectivity, the main limiting factors for its current use in these countries is inadequate access to data networks, and the availability and cost of suitable smartphones. The global coverage of such networks is likely to increase over the next few years, eliminating the problems.

Another issue with this method involves the battery life of mobile phones. The team has suggested that users should buy solar-powered mobile phone chargers, which are now widely available at a low cost to solve the problem.

EpiCollect was developed for Android due to its open-source nature and the provision of suitable multiple phones by different network providers. The team behind *EpiCollect* envision that versions will soon be provided for other Smartphone operating systems such as Windows mobile and Palm webOS. The iPhone now has a version of EpiCollect available for it.

Technological expertise required:
Staff skills required: Must be able to use the software to create a project, mobile devices and their operating system. They must also be able to help fieldworkers with the two-way communication between users and the project databases.

Student skills required: Must be able to use the mobile devices and understand how to transfer the information they gather to their project database.

Also see: Aanensen et al. (2009).

Case Study 11: Situ8 for Creation of Geo-located Content (Elizabeth Fitzgerald)

Institution: Open University Institute of Educational Technology

Keywords: *Situ8*, citizen science, insects

Aim: To gather field data and observations using *Situ8* software

Situ8 is a tool designed to enable the delivery and creation of geo-located user-generated content, referred to as Media Objects (MOs). It can be used for both formal and informal learning, citizen science and collection of fieldwork data. It is especially useful for carrying out initial field observations and capturing site descriptions. It can also be used to upload and share fieldwork data (multimedia, text and numerical data). Users can browse or download existing MOs and filter them according to date, author, subject and/or tag.

A field example

Situ8 has been used to count the number of individuals from different insect groups that act as pollinators on flowering plants and also the wind speed and ambient air temperature. This means that students can make hypotheses about how these variables affect each other and use *Situ8* as a fieldwork data-gathering tool.

Quantitative data and text entries uploaded to a student's *Situ8* account can be downloaded directly and shared with other *Situ8* users and also made publically available, depending on what settings the user chooses.

These data are saved into CSV (comma separated variable) format to ensure easy import into other software packages and enable the data collected in *Situ8* to be analysed back in the laboratory or used for assessed coursework. Likewise, photographs, videos and audio clips can all be shared, thus allowing a truly collaborative field-work experience.

The insect pollinator exercise could easily be adapted for use elsewhere, for example to investigate how populations of insect pollinators differ between different habitat types, or to find out which other environmental variables affect insect pollination.

For more information go to SitU8 (2013).

Case Study 12: Using GPS and ArcGIS for Environmental Science Fieldwork at Lawrence University (Meg Stewart, Jeffrey Clark, Jeremy Donald and Keri VanCamp)

Institution: Lawrence University, Wisconsin, USA

Keywords: Mobile technology, fieldwork, teaching, learning, tablet PCs, GPS, ArcGIS

Aim: To record, capture and analyse field data with ArcGIS.

The Introduction to Environmental Science course taught at Lawrence University acts as a gateway course into the Environmental Studies major. The course teaches 60 students (approximately), divided into three-hour laboratory sections of 20 students. The course's main learning outcome was to achieve proficiency in the collection and analysis of data, especially geospatial data. This time period aims to fulfil the course's laboratory-science requirement.

The laboratory component of the course consisted of two sets of paired exercises. Two of the exercises focused on the development of relevant technology skills, such as using ArcGIS and a GPS receiver to collect and analyse field data. Figure 4.2 shows students preparing for fieldwork with a tablet PC and GPS receiver attached to the student's cap. The second sets of paired exercises are application-type exercises, in which students must use various tools to analyse the environmental data. These application exercises

Fig. 4.2 Students learning how to use GIS technology in the field. (Image courtesy of Stewart, Clark, Donald and VanCamp)

reinforce the technology skills exercises and engage the students to answer relevant questions in environmental science.

In the first pair of exercises students are taught how to collect, add attributes to, and analyse point data. These technological skills link into the application exercises that require students to map water quality parameters through an urban stream setting; particularly focusing on spatial analysis tools including interpolation and contouring. These skills are then used to map lead contamination of urban soils. Figure 4.3 shows an interpolated map of soil lead concentrations around Appleton prepared by students. Students then formulate hypotheses on the sources of contamination and create suggestions for mitigation.

The tablet PCs used in these activities served primarily as location-aware digital field notebooks. Students mark a sample location using a GPS and an aerial photograph, and add in situ attributes. Students measured temperature, pH, dissolved oxygen and conductivity in the field, and enter these data into GIS. The geo-database and relevant attribute fields must be set up in advance. Students can then see spatial trends (via simple colour coding of temperature) and ask questions about those trends.

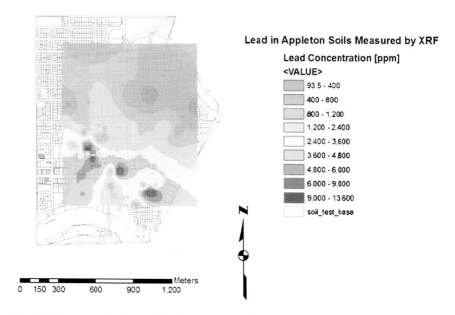

Fig. 4.3 An example of a soil lead concentration interpolated map around Appleton. (Image courtesy of Stewart, Clark, Donald and VanCamp)

Technology required:
Hardware used: HP tablet PCs, GlobalSat USB/Bluetooth GPS receiver
Software used: ESRI's ArcGIS, Microsoft Office
Approximate cost of technology: HP tablet PC: £600.
GlobalSat USB/Bluetooth GPS receiver: £25

How much time was taken to develop this and implement the method?
It took 5 h per week to prepare the laboratory for the students taking the course.

What evidence is there of the effectiveness of this activity?
To evaluate the achievement of the learning goals in the Introduction to Environmental Science course, pre- and post-course questionnaires were used. The questionnaires revealed that the students felt that the course improved their general scientific and quantitative skills; such as creating conceptual models, preparing laboratory reports, interpreting experimental results, designing experiments and using graphs to analyse their data. Students also significantly improved their perceived skills and understanding of geospatial analysis, including making and interpreting maps, as well as knowing the functions, uses and limitations of GIS.

The evaluation of the students' work also supports the questionnaire's findings. The first exam covered course content and the first two laboratories.

The class average for this was around 65 %. After two more laboratories the class performance increased to an average of 78 %. Scores on laboratories also generally improved from lower 70s to higher 80s; therefore the findings show that overall students achieved the main learning outcome, as they improved their proficiency in collecting and analysing environmental data and geospatial patterns.

Pitfalls/Problems/Limitations:
The implementation of tablet PCs was found to be very successful and worthwhile during the course, but they also had a number of problems. Problems can come from the hardware, software and the novice user. Students were provided with step-by-step instructions, many problems did not occur until all 20 students were using the mobile devices at once. When going from laptop mode to tablet mode the screen orientation changes which hides some of the tool bars and changes the display scale. If students don't know how to reset the orientation, then they cannot find the proper tools. An instructor or well-trained teacher assistant can easily rectify these problems in the field.

The technology appeals to students, but the students are very intolerant of glitches with the various technologies, even when the problem is caused by their own actions. Many students appreciated having a fully functional computer with the standard Microsoft Office software, but few took advantage of the capabilities. It seems that most students are still more comfortable with using paper and pencils in the field. Unless the tablet was required for a particular task, the students chose to leave them behind. Some students left them because of their bulk, difficulty in viewing the screen outdoors and fear of breakage; however, over the course of three years a student has yet to damage one in the field.

Technological expertise required:
Staff skills required: Must be able to use the technology, and teach and support the students when they are using the technology.

Student skills required: Require basic computer skills as these will be used on the tablet PCs, such as Microsoft Word. The tablets act as digital field notebooks.

Also see: Stewart et al. (2011).

4.6 Summary

This chapter highlights the potential to develop more robust opportunities to create and distribute digital information through a range of alternative sources. Students are now able to consume and record information (see case studies 7 and 8) in the field through the creation of a digital field notebook. A variety of apps are considered and case studies 9 and 10 offer worked examples of applying *Epicollect* as successful platform to collect field data.

References

Aanensen DM, Huntley DM, Feil EJ, al-Own F, Spratt BG (2009) EpiCollect: linking smartphones to web applications for epidemiology, ecology and community data collection. *PLoS ONE* 4 (9):e6968. doi:10.1371/journal.pone.0006968

Beddall-Hill NL, Jabbar A, Al Shehri S (2011) Social mobile devices as tools for qualitative research in education: iPhones and iPads in ethnography, interviewing and design-based research. J Res Cent Educat Technol 7(1):67–89

Edge (2014) Evolutionarily distinct and globally endangered species website. Retrieved from http://www.edgeofexistence.org

Epicollect (2014) Epicollect application website. Retrieved from www.epicollect.net

France D, Whalley WB, Mauchline A (2013) Using mobile devices to enhance undergraduate field research. CUR Q. 34 (2):38–42 (Council on Undergraduate Research, Washington)

Friedland G, Sommer R (2010) Cybercasting the joint: on the privacy implications of geo-tagging. HotSec'10 proceedings of the 5th USENIX conference on hot topics in security, Article 2. USENIX Association Berkeley, USA, 8 p

Harrison C (2011) The Great Yew Hunt Project Website. Retrieved from: epicollectserver.appspot.com/project.html?name = YewHunting

Jackson R (2012) Retrieved from http://blogs.reading.ac.uk/engage-in-teaching-and-learning/2012/09/03/heading-to-the-arctic-to-teach-students-about-the-wonderful-world-of-extreme-microbes-by-dr-rob-jackson/

Luo J, Joshi D, Yu J, Gallagher A, Gallagher A (2011) Geotagging in multimedia and computer vision: a survey. Multimedia Tools Appl 51:187–211

Middleton A (2013) Digital voices. Retrieved from http://melsig.shu.ac.uk/wp-content/uploads/2013/01/Digital-Voices.pdf

SitU8 (2013) Online platform to innovate places with digital information. Retrieved from http://www.situ8.org

Stewart M, Clark J, Donald J, Van Camp K (2011) The educational potential of mobile computing in the field. Retrieved from https://www.educause.edu/ero/article/educational-potential-mobile-computing-field

Traxler J, Wishart J (eds) (2011) Making mobile learning work: case studies of practice. ESCalate HEA Subject Centre for Education, Bristol

Welsh KE, France D (2012) Spotlight on…smartphones and fieldwork. Geography 97(1):47–51

Welsh KE, France D, Whalley WB, Park JR (2012) Geotagging photographs in student fieldwork. J Geography High Educ 36(3):469–480

Chapter 5
Utilising Video in Fieldwork

Abstract In this chapter we consider various applications of video to enhance fieldwork learning including the development of Podcasts and Vidcasts. We also include some advice on using iPads and other mobile devices for video recording for educational purposes. A number of case studies are presented to exemplify the use and application of video.

Keywords Video · Podcast · Vidcast · Digital video technology · Virtual learning

5.1 Video Recording

At their introduction, iPads were thought of as devices to consume media, in particular, of video as well as for web browsing. However, it rapidly became apparent that video production was well within the capability of iPads and iPhones.

Case Study 13: The Use of Video in Field Teaching (Julian Park)

> **Institution:** University of Reading, UK
>
> **Keywords:** Environmental science, video diary, iPad Mini, Flip Video
>
> **Aim:** For students to produce a video diary of fieldwork activities.
>
> There is a long tradition of using either still or moving images to capture elements of fieldwork. Historically, video capture entailed the use of bulky and expensive equipment, which was often not well-suited to the outdoor environments in which it was used. Further, the editing of film, once back at the University was a specialist and time-consuming process. One of the assignments related to our field courses (Environmental Science) was for students to produce a video diary of the field activities during the week and for students to reflect on their learning experience. The finished video could be shared with peers after the field course, who had experienced different field

environments or for those students that had yet to go on field week. Some of the material produced has also been used for promotional activities.

Video technology has moved on considerably in the intervening period, both in relation to the technology available and the subsequent editing of material post-capture. Video-recorders have become simpler to operate, are generally more environment-proof and less easily damaged in adverse conditions. Indeed many of the tablet computers and mobile phones we carry have video capability.

This case study sets out to encourage the use of video in a range of field contexts:

1. The recording of short clips, for instance, explaining how the positioning of a transect was decided, which can then be replayed, perhaps as part of the subsequent evening's follow-up activity within the field course itself.
2. The recording of key aspects of a field site in one year, to be shown to students in advance of a subsequent field visit to that site, so as to enable orientation and feed-forward re assignments, thus potentially saving valuable time in the field.
3. The use of video to produce reflective field diaries either in groups or individual reflective diaries. These can form a component of the assessment but also provide a resource for other students.

Technology required:
This now involves cheap and accessible technology. Small hand held devices such as a *Flip* video camera (similar to other devices such as the Kodak *PlayTouch*, which is waterproof), although now discontinued, provides adequate quality pictures, although it is probably advisable to invest in slightly more expensive recorders with an external microphone and the ability to zoom. Footage can be edited on computers, if necessary, via a range of software packages, and can usually be replayed through most audio-visual equipment. Tablet computers, such as iPads, and smart phones have integral video cameras with a range of apps for video editing such as *iMovie* and *Splice* allowing the recording, editing and playback to be done on one device. The key point is that the technology is usually cheap and accessible. Further, it is often a technology and medium with which students are familiar.

Approximate cost of technology per unit:

- Flip video £80 or small dedicated camcorder £250
- DSLR with HD video capability starting at £380
- Editing software £50 (but free with app such as *Splice*)
- iPad Mini £270 with free or inexpensive apps
- Smartphone with video camera

How time intensive is this to set up and implement?
It is important to ensure that students are familiar with the equipment being used. This is best undertaken in the classroom prior to fieldwork. This also provides the opportunity for students to complete a formative assignment, perhaps linked to some form of peer assessment to ensure good practice is shared.

Actual set up time when in the field is minimal. Editing is the most time consuming element, the requirement for which will depend on the context in which the video recording is being used. If appropriate, pre-use 'story boards' can help focus video footage collection, and should reduce editing time.

What evidence is there of the effectiveness of this activity?
There is an increasing body of evidence outlining the benefits of video in teaching and learning, see Asset (2014) for an overview. Experience suggests that students enjoy undertaking video assessments and that, depending on the type of assessment set, it provides a mechanism for encouraging reflection after the field activity has been completed, as well as encouraging thought prior to the fieldwork about what can usefully be recorded.

Pitfalls/Problems/Limitations:

- Ensure students understand how the technology works, which means familiarising students with the technology prior to the field experience.
- If you are operating in very wet conditions then some care is required with standard equipment. It is possible to purchase fully waterproof technology but this can add considerably to the costs.
- Most recorders available either operate via batteries or rechargeable battery packs. Obviously it is important that spares are available and that the packs are fully charged.
- Ensure appropriate cables are available for download, although some devices such as the flip have their own built in USB port.
- Some individuals will not want to be recorded on video, and these wishes need to be taken into account.
- If wider use is going to be made of videos then permission of participants will be required.
- Successful video requires good sound capture as much as video. A note on this is provided in a subsequent section.

Technological expertise required:
Minimal expertise required by staff or students, although familiarisation prior to the actual field experience is essential. Software editing requires some practice but is increasingly simple. Complex editing can be quite time-consuming but for many purposes may not be necessary at all.

Key tips:

- Make sure students are familiar with equipment before going into the field
- Prior storyboarding can reduce editing time and ensure appropriate footage is gathered
- Make sure equipment is functioning OK before going into the field
- Where using group work, try and convene groups with a mixture of "video" abilities
- Consider how to share the videos e.g. set up a private YouTube channel
- If video is going to form part of the assessment, ensure you have clear assessment criteria and communicate these to the students beforehand.

Also see: Asset (2014) and France and Wakefield (2011)

Case Study 14: Applying Digital Video Technology in Physical Geography Fieldwork (Ian Fuller)

Institution: Massey University, NZ

Keywords: Digital video, physical geography, active learning

Aim: To incorporate video diaries into residential fieldwork to enrich and enhance the student learning experience.

Teaching Method:
The field course takes the form of a 7-day road trip, which encompassed a range of environments, topics and issues as part of a 3rd year undergraduate paper in applied field geomorphology running around the highly erodible terrains of the New Zealand's North Island's East Coast and through its volcanic Central Plateau regions. The intention and genre of the trip is observational, rather than hands-on measurement of process, with an emphasis on the relationship between evaluation of process and critical assessment of landscape management approaches within a broad spectrum of geomorphic topics (e.g. landslides, flooding, coastal erosion, volcanic hazards).

Hands-on technical skills are taught using field experimentation investigating fluvial processes and process geomorphology in an alpine setting. The former were studied through a series of short, 2–3 h long field experiments using a stream local to the University campus in Palmerston North, New Zealand. These experiments run in the normal course of the teaching day and deployed equipment including electromagnetic flow metres, a total station, cone penetrometer and standard grain size sampling tools (sieves and pebble template). Experiments were designed to measure stream flow and resistance, bed sediment characteristics, pool-riffle morphology and channel morphodynamics. Process geomorphology was studied as part of an 8-day residential fieldtrip at Fox Glacier, South Island New Zealand, a site remote from the University and in a highly contrasting, alpine environment. In addition to equipment described above, field experimentation used a high-precision Global Positioning System using Real Time Kinematic survey (RTK-dGPS) to measure glacier velocity using surveyed ablation stakes. Cross-profiles of the Fox Glacier sandur and its sediment assemblages were also measured (see Fig. 5.1), as were flow properties in the proglacial stream. In both courses, students work in self-selected groups of c.4 and had the responsibility of setting up and running the equipment to derive data, which they processed and interpreted in their own time or, during the evenings of the residential fieldtrip. The intention of the fieldwork in these courses was to provide detailed, hands-on technical experience measuring key parameters in rivers and alpine landscapes, providing an introductory research experience (see Fuller and France 2015).

Please indicate what technology is required:
Digital camera with video capability/smartphones/tablets/flip-videos. Tripods

Approximate cost of technology:
Variable, £50 up for digital camera

Please estimate how much time was taken to develop this and implement the method:
Initially after conception a few hours i.e. quick

What evidence is there of the effectiveness of this activity?
This has been comprehensively evaluated over a number of years from 2010 onwards when this intervention was introduced. Strategic incorporation of video diaries as described in Fuller and France (in press) and (2015) can enrich the learning space and enhance the learning experience. Incorporating digital video as part of field experiments in process geomorphology proved enjoyable and improved understanding of methods in particular, as well as the processes being studied.

Please outline any pitfalls/problems/limitations:
Inevitably, not all students appreciated the opportunity for participation and learning facilitated by the introduction of digital video into these field courses. A very small minority expressed this in terms of hating using the video camera and perceived it as a waste of time (cf. Fuller and France 2015). To be fair, this might be true of the best students, who probably benefit least from its use, because they are already engaged and stimulated to learn. Nevertheless, the overwhelming majority believed that producing a digital video was a valuable and helpful exercise (Fuller and France 2015 and in review). Interestingly, students did not necessarily appear to recognise the links between their video production and other written assessments

Fig. 5.1 Students surveying the glacial foreland at Fox glacier, New Zealand

(e.g. report). The deeper, holistic learning connections between the use of video and later assignments, which build on knowledge generated, were not always perceived by students (Fuller and France 2015). However, this may also reflect the need for clearer instruction and communication of the linkages between assessments by staff. Links between various components in a course cannot be assumed and need to be clear.

Technological expertise:
Staff skills required:
Ability to use a camera, have some understanding of the basics of video story telling and the application of a storyboard prior to filming and basic video editing using free software e.g. *Windows moviemaker*.
Student skills required:
Ability to use a camera, have some understanding of the basics of video story telling and the application of a storyboard prior to filming and basic video editing using free software e.g. *Windows moviemaker*.

> **Key tips or advice to others:**
> Practise before they go in the field and read some appropriate literature on the production and application of video to fieldwork. e.g. France and Wakefield (2011).
>
> **Useful references:**
> Fuller and France (2014)

The next case study, by Susan Mains and colleagues, shows how the relatively simple technologies of iPads and movie apps, can be brought into several days of a field class.

Case Study 15: Promoting Active Learning and Mixed Methodologies Using *Fotobabble* and *IMovie* in the Scottish Highlands (Susan P. Mains, Lorraine van Blerk and Jade Catterson)

> **Institution:** University of Dundee
>
> **Keywords:** active learning, visual methods, landscape interpretation, mixed methods, qualitative research, *Fotobabble, iMovie*.
>
> **Aim:** To embed mobile technologies into human geography fieldwork.
>
> **Learning Context:**
> *Fotobabble* and *iMovie* were used to support pedagogical themes and integrated into students' assessments (Human Geography component) of the residential Aviemore field trip based in the Scottish Highlands. This Level 2, core module involved a fairly large group (~ 85 students) so the class was divided in two.
>
> **Learning Objectives:**
> The goal of the Aviemore field trip is to promote a greater understanding of the Physical and Human Geographies of the Scottish Highlands, and to develop methodological approaches that facilitate more nuanced and flexible research skills.
>
> **Student Activities:**
> Human Geography-Day 1 (*Fotobabble* use led by Lorraine van Blerk)
>   Day 1 began with a preparatory discussion about the goals of the assignment and ethical research, including a brief discussion of possible approaches that could be taken (including *Fotobabble* and the iPads). The class organised themselves into groups of 5–6 students.
>   The goal of Day 1 was to examine rural life and use a range of themes—e.g., youth culture in a rural community—to highlight significant issues. The key method to be used was interviewing and then this could be developed through other approaches, such as a small survey, photography and/or field

observations. Students were to work in groups and the data collected were to be presented on a poster, accompanied by a 10 min presentation to the other groups and teaching staff, followed by 5 min of questions. By bringing our own 'field printer,' we were able to print out photographs (with a guideline maximum of 4 pictures per group).

Each group then refined their chosen topic and were dropped off in a range of nearby villages for approximately four hours. The students were then collected and came back to the hostel to compile their data and prepare their group posters and presentations (which took place in the evening). *Fotobabbles* were integrated into these reflective posters in conjunction with other elements, such as research questions, village maps and a range of brochures or literature that students had collected during the day.

Assessment: The grading was focused on the students' ability to answer their research questions and engage with their central theme—via their poster, verbal presentation and through a demonstration of collaborative work. We compiled our feedback and grades at the end of each day, which were then returned to the students prior to work commencing the following day.

Human Geography—Day 2 (*iMovie* led by Susan P. Mains)

Day 2 began with feedback on the previous assignment, followed by a preparatory discussion about the goals of the next assignment and how these could build on skills learned from Day 1. This presentation included a discussion of landscape interpretation in fieldwork and how this could be represented through *iMovie* on the iPads. Each group then refined their assigned topic in relation to Inverness. To give some contextual information prior to commencement of the student's own projects, we arrived as a group in the city and we led the students on a brief walking tour along the River Ness into the city centre. The students then spent approximately 3½ h collecting data and filming. We then returned to Aviemore, where the students continued to analyse and edit their data and prepare their films and presentations (which also took place in the evening).

The goal of Day 2 was to examine urban landscapes in Inverness and use a range of themes—e.g., heritage and tourism—to highlight significant issues. The key methods to be used were landscape interpretation and film documentation, in a similar manner to the previous day, this could be developed through additional approaches, such as interviews, media analysis of tourism materials, photographs and/or field observations. To provide some continuity, students worked in the same groups as the day before, and the data collected was to be presented as a short film (up to 3½ min long), accompanied by a 7½ min presentation to the other groups and teaching staff, followed by 5 min of questions.

At the outset we intended each group to produce a 1½ min video, but to allow all students greater participation while having a sense of ownership of the output, a slightly longer video seemed more appropriate. This slightly longer length also provided a greater opportunity to include a more diverse range of information and reflection. *iMovie* was the preferred software, students could film using the camera on the iPad, and/or film directly via *iMovie* (to reduce the editing process).

Assessment: Grading was again focused on the students' ability to answer their research questions and engage with their central theme via their film, verbal presentation and through a demonstration of collaborative work.

As a general guide, the students were advised to provide a brief (1–1½ min) verbal introduction to their topic, followed by their film, then a (5–6 min) discussion of the methods used during the fieldwork and the opportunities/challenges faced while collecting their data. Clarity and good organisation were key components of the exercise. Students were encouraged to use the films as a stand-alone informational tool, to avoid repeating what had just been shown on-screen.

What technology is required?
Hardware: iPads: to take photographs, record videos, and edit videos
  Printer: for producing hard copies of photographs to incorporate into posters
  Mobile projector: to enable viewing of short films (and to present introductory information in relation to software being used)
  Software: *Fotobabble*: for recording and producing annotated photographs of field sites and places/items of interest
  iPhoto/photo booth: for recording and producing photographs of field sites and places/items of interest
  iMovie: for recording and producing short films exploring key themes of the field trip

Cost of technology:
iPads (x8) & Care Plan: £3,312; Tough Cases (to protect iPads x8): £232
  Mobile Projector: £204; Wireless Printer: £59
  Software (*iMovie* & *iPhoto* x8 (*Fotobabble* is free)): £48

How much time was taken to develop this and implement the method?
Self taught tutorial with the software to produce annotated photographs and short videos prior to departure/on arrival at our field site. Having more hands-on time ahead of the trip would probably be useful for getting more familiar with different features, but this is not absolutely necessary.

What evidence is there of the effectiveness of this activity?
Feedback from students during and following the field trip was very positive. Students also showed a lot of creativity and produced some very high quality materials. An unintended benefit came in the form of student engagement: the video component helped maintain audience attention during presentations—class members seemed quite keen to see how their colleagues' films had taken shape.

This exercise was a learning experience for staff and students. For example, students highlighted new features e.g. editing their projects. We found the experience a much more collaborative and interactive learning process.

Pitfalls/Problems/Limitations:
We did have some issues with limited Internet access, but circumvented many of these by using *iMovie*, which was already installed in the iPads. We did not ask students to bring their own laptops: This was for security and equal opportunity reasons, and for students to concentrate on the apps provided, but in the future we could try to digitise the posters and attach them to the university VLE or a blog associated with the field course.

The students seemed to engage more actively with *iMovie* rather than *Fotobabble*—we think the moving image media seemed quite exciting and appealed to aspiring new film directors! We could introduce *Fotobabble* earlier in the semester during workshops, so that students have more time to think about different ways they could creatively use this app.

Technological expertise required:
Staff skills required: A willingness to engage with new technology—the touch screens on the iPads are quite intuitive and fairly easy to navigate and the software were relatively straightforward for novice users—so no specific expertise is required. Enthusiasm and a sense of humour do help, particularly in relation to unintentionally entertaining video clips.

Student skills required: Same as above, indeed many students were already familiar with the general format of the iPads through previous use of smartphones and tablets. A willingness to work in groups and collaborate on the development of field information is beneficial, as well as the ability to take on different roles during the research process (e.g., taking notes, then filming or being in front of the camera, assisting with story development and/or editing).

Key tips or advice to others:
Just do it! The best way to learn the most effective ways to use the technology is to put it into practice, rather than worrying too much about perfecting your knowledge of the hardware/software. Students appreciate efforts to try something new and have them being actively part of that process—and then it becomes a joint learning journey and confidence builder.

Set a guideline limit for the number of photographs that students need for posters and a time limit for videos. We found 4–5 photographs and 3½ min max.

Provide an ethical usage guide for students using iPads—particularly relevant in relation to appropriate use/informed consent in relation to photography, filming and accessing the Internet.

> Encourage students to keep thinking about how they are addressing the key topics being explored and what interesting things they have discovered, and not just how to use the iPads.

5.2 Podcasting

We now devote some pedagogic reflection on the use of Podcasting (a term that now includes video and has tended to replace terms such as Vodcast).

Podcasts and podcasting are terms that have emerged since 2004 when they were used to describe 'portable listening' to audio-blogs Hammersley (2004). Users initially could subscribe to and download audio files directly from a website to their portable device; an mp3 player such as an iPod. While originally used for entertainment and information (Salmon and Edirisingha 2008) podcasting has been used for a variety of applications in higher education. The drive behind this can be attributed to pedagogic necessity, technological innovation, funding and the apparent inadequacy of traditional e-learning (Traxler 2009). "The increased confidence expressed by students that appears to relate to the multi-modality of the materials is particularly noteworthy when weighing up the value of the podcasts as a means of improving the students' learning experiences" (Jarvis and Dickie 2010).

Originally coined from iPod and broadcasting (Kemp et al. 2012), podcasting later became synonymous with vodcasting; which includes a video element. The additional video element allows students to view complex demonstrations repeatedly and in their own time, potentially via a VLE or webpage. One of the main benefits of podcasting is the availability of resources for students who benefit from audio-visual learning styles, allowing all students to connect with VLE materials and to be motivated by hearing the tutor's voice (Lee and Tynon 2008). In particular, podcasts are useful for recapping 'forgotten' material, such as elements that were discussed in a previous academic year, and they can be used as revision aids (Jarvis and Dickie 2010).

Another benefit of podcasting is the minimal financial outlay for equipment (Salmon and Edirisingha 2008). Portable devices can be bought for a reasonable cost and frequently students have their own compatible devices. Widespread ownership means it is often unnecessary to train students how to use the hardware. Additionally, creation of podcasts is also low cost, requiring only a computer, camera, microphone and some inexpensive software. Arguably the most obvious application of using podcasts in higher education is to provide students with a digital copy of lectures. However, in recent years practitioners have shown increasing innovation when using podcasts for teaching and learning in general.

One of the most extreme applications of podcasting have been to completely replace lectures with podcasts and use the time for practical laboratories (Gannod et al. 2008). Foothgill (2008) demonstrated how podcasts were used to deliver a full engineering module via online learning in a VLE rather than conventional lectures, citing that students liked being able to listen to podcasts in their own time, effectively being on demand. His findings suggested that students were much more positive and accepting of this method of delivery than would be expected. Jarvis and Dickie (2010), however, found that the student preference was to use these resources to support rather than replace conventional methods, as they did not want to lose the student-teacher interaction.

Case studies that use podcasting for demonstration purposes have shown positive engagement and feedback from students across a number of disciplines (Lee and Tynon 2008; Cox et al. 2008; Cake 2006; Heilesen 2010). Jenkins and Lonsdale (2008) outlined how podcasts can be made by students as a form of digital storytelling. This cross-disciplinary study at the University of Gloucestershire encompassed staff and students from Environmental Studies, Landscape Design, Accountancy, Sports Development and Leisure & Tourism courses, demonstrating the versatility of digital storytelling via podcasting within an academic setting. As well as digital storytelling, assessments through podcasting have been successful and include outputs such as constructed radio programmes (Kemp et al. 2012) and presentation of report sections—results or methodology (as seen at the University of Chester). This range of outputs increases the skills that graduates have, specifically digital, information and academic literacies. Students have also demonstrated increased confidence with using multimodal materials in podcast creation (Jarvis and Dickie 2010). The case studies included in this section illustrate some of the key benefits and perceived pitfalls of introducing podcasting to a degree module. These examples are transferrable across a range of subjects, focusing on fieldwork reflection and research development through podcasting.

Case Study 16: Podcasts to Support Student Learning in GEES (Ming Nie)

Institution: University of Leicester, UK

Keywords: Podcasts, GEES subjects, IMPALA project, virtual, CD-ROM, DVD.

Aim: To evaluate subject specific benefits of podcasts as a learning tool for students.

IMPALA (Informal Mobile Podcasting And Learning Adaptation) 2, focused on the GEES subjects at the University of Leicester to better understand the use of podcasts and to evaluate and assess how the students engage with this new style of learning and what problems if any occurred. A further five institutions were also included in the IMAPALA 2 study (Nottingham, Leeds, Sussex, Gloucestershire and Kingston).

The IMPALA 2 project started by giving GEES subject teachers a chance to attend two workshops, where they could share ideas on the use of podcasts in their lectures. From this, it was identified that different academics prefer to use podcasts in a variety of approaches, as follows:

- Podcast to provide additional support and information about the subject
- Podcasts to act as a virtual field guide
- Podcasts to explain field equipment and how to use it
- To enable students to create digital stories for assessment purposes
- To provide lecture summaries and feedback for students

One main topic that academics in the GEES subject areas used podcasts for was in supporting field based learning. Due to the wide availability of low cost handheld devices, more off site learning is becoming commonplace. The idea of podcasting in fieldwork is to take snippets of the classroom away from the campus and direct it into the hands of the students in the field. Students listened to instructions provided via the audio and completed their fieldwork in their own time.

Other uses of podcasts in fieldwork have been via CD-ROM or DVD while on the coach/bus to the fieldwork site. Delivering instructions on how to set up and use equipment, the objectives of their tasks and an overview of the fieldwork site. Using the transit time wisely and thus saving wasted time in the field.

Finally, libraries of geographical techniques were compiled by Leicester University, so that students could access them in their own time. Either for refreshing their own skills for independent research or for revision.

Other uses of podcasts:

- To improve learner engagement and motivation
- Offering a new dimension to learning with added flexibility
- Provide effective feedback for students
- To help students understand the troublesome concepts relating to the subject.

Technology required:
The technology required ultimately depends on the scale of the podcast. Most podcasts are basic and include video and audio, the technology required for this:

- A computer with video software (Microsoft *Moviemaker* live or similar products)
- Digital resources such as audio recording equipment, video equipment and digital photography
- Devices for the podcast to be placed onto- intranet, CD-ROM, DVD, MP3, mobile phone device or others such as tablet PC's. However some podcast makers make their podcasts more professional through the use of specialised video software such as, Adobe *Flash* or *Premier Pro* CS5.5. NB, Apple's iOS will not run Flash although it can be run via the *Puffin Web browser* on a puffin server.

Approximate cost of technology:
The cost of technology varies with use (c. 2011):

- PC from £250; Digital resources start from £80
- CD-ROM, DVD are relatively cheap to attain
- MP3- £60 to £200 (depending on make, model and storage capacity)
- Tablet PC's—£300 up; Adobe *Premier Pro* CS5.5—£539

What evidence is there of the effectiveness of this activity?
Key findings from student interviews on the whole were very positive when it came to using podcasts as an aid to learning. Students were keen to point out that the podcasts helped them save time in the field and to set up and use equipment correctly. Staff noticed an improvement in attention and participation from the students. Students were noted as saying they enjoyed the added learning experience offered through the use of podcasts and finally they appreciated the feedback aspect of the podcast.

Staff interviews also showed that they were keen to build up a digital library of podcasts that could be used repeatedly within the university. They were also keen to build up a communal library where academics could share their podcasts with one another. The idea of a communal library is something that academics are considering and it could be implemented within the near future. The latter seems to be one for the immediate future.

Technological expertise required:
Staff skills required: some knowledge of computer systems and *Microsoft Moviemaker* for podcasts. The more skill attained the more professional the podcast will be.

Student skills required: very little—As long as they can access a computer or work the selected device.

Also see: Nie (2008), IMPALA (2012)

Case Study 17: Acknowledging the 'Forgotten' and the 'Unknown': The Role of Video Podcasting for Supporting Field-Based Learning (Claire Jarvis and Jennifer Dickie)

Institution: University of Leicester, UK

Keywords: Podcast, teaching, learning, pedagogy, m-learning

Aim: Using a Podcast library to support the teaching of field methods.
The teaching method involves the construction of a podcast library, which was used to instruct students on how to use specific field equipment relative to the science of soil properties, water quality, atmospheric variables, and field surveying. The podcasts provide information and operational demonstrations for a range of field equipment. The podcasts all relate to soil

properties, water quality, atmospheric variables, field surveying and mobile technology. The equipment and technology referred to in the podcasts fall into three categories:

- A familiar piece of equipment which has usage issues by both students and staff each year.
- Equipment that is not considered to be core and is unlikely to be covered or even introduced within classes. Although the equipment isn't deemed as core, students in specific dissertations commonly use it.
- New equipment that students or staff are uncomfortable with. In this case the podcast explains and demonstrates the new equipment while supporting change.

The podcasts can be downloaded or viewed at Impala (2012). The content is categorised into the relevant subject areas. The podcasts were enabled in both high and low resolution graphics formats. The podcast library was also uploaded onto a variety of mobile equipment, in the format best suited to each device. The podcasts were first designed for students to use in the field using iPods. This was because textbooks were not available out in the field and practitioners were not always at hand.

Technology required:
The technology required is podcasting equipment capable of recording audio-visual footage. The podcast library explains about various field equipment and techniques. By using the podcasts to support the fieldwork, this enables the lecturers to spend more time with the students on the 'intellectual challenges', rather than the spending time explaining and demonstrating how to use the equipment in the field.

How much time was taken to develop this and implement the method?
The podcast library was initially trialled over 6 months.

What evidence is there of the effectiveness of this activity?
Feedback from staff and students revealed that the podcast library had many advantages. The podcasts were supported and their usage was much higher than expected, as students used them in a number of scenarios. The podcasts were particularly successful due to the amount of students who prefer visual learning, the podcasts succeeded as learning and preparatory tools to fieldwork. If was also found that students use the library as revision tools when it came to field equipment and techniques. The students felt that the library had boosted their confidence in fieldwork.

Pitfalls/Problems/Limitations:
The method has no apparent limitations. It does not aim to replace teacher demonstrations of traditional teaching methods. Instead the podcast library aims to act as supporting materials preparing students for fieldwork.

Technological expertise required:
Staff skills required: Must be able to produce the podcasts using the relevant technology (such as cameras and editing software), and must be able to upload them so that students and staff can view them at any time.

Student skills required: No skills required. They learn by watching the videos that the teachers show them or direct them to.

Also see: Jarvis and Dickie (2009); Impala (2012)

The next case study may look like an old-fashioned lecture delivery. Yet delivery of instructional material may be necessary in all sorts of places other than lecture theatres. A field centre is just one. Students might want to make reports of what they have done and the technology now allows this whether from desk or laptops (as here) using the screen capture application *Camtasia*. This is a semi-professional application and, like its cheaper kin *Jing*, does not exist on tablets. There are possible alternatives: *Skitch*, covered elsewhere. But the important thing is that there are apps which allow creativity for students to explore.

Case Study 18: Virtual Learning: Delivering Lectures Using Screencasting and Podcasting Technology (Sandy Winterbottom)

Institution: University of Stirling, UK

Keywords: Virtual learning, VLE, Intranet, Screencasting, podcast, *Camtasia, PowerPoint*

Aim: To assess the usefulness of downloadable and portable lectures to students.

Screencasts are a digital recording of what you see on your computer screen, often including imagery and the optional extra of a webcam recording to support the imagery. Screencasting is used mainly to record *PowerPoint* presentations through software as *Screencast-o-matic, Jing pro, Microsoft Producer* or *Camtasia Studio*.

This study was conducted with a second-year undergraduate environmental science module in Earth and Landscape Evolution. The first 8 of the 20 lectures were delivered using screencasts and podcasts. This was undertaken to determine the best delivery of information to students and to see which was the most popular.

Podcasting creates a media file that contains video and/or audio and can be distributed over the web or downloaded onto a PC and then transferred to any selected device; usually a MP3 player or mobile phone. Screencasts have an advantage over standard audio recordings as they can provide imagery and diagrams. This is very important for a subject like environmental science, as it is a very visually-rich subject.

The first 8 lectures were recorded prior to the start of the module using *Camtasia Studio*. It was believed that they should be produced in both high and low resolution and in a variety of formats, from MP3 to Mpeg4. This was so the varying devices the screencasts/podcasts will eventually be shown on could support the file, and that the students could decide which format was best for their device. Once completed the audio file was exported to *GarageBand*. This was to enhance the podcast audio and convert it into Mpeg4, which allowed the podcast to have chapters and images to improve navigation through the materials.

The files were then uploaded to the online module space, which the students regularly access throughout the term via the University's Intranet. On the online module space there was an introductory page containing full instructions for viewing and downloading the e-lecture with accompanying instructions on what to do if problems arose (however, no students reported any problems). Two lectures were placed online each week with an automatic email being sent to the students to inform them of the new material.

The lectures were uploaded to the module space in all formats and resolutions including a Zip file and a PDF of the *PowerPoint* slides. This continued until all 8 lectures had finished. This system of delivering lectures online was on the whole a very positive one for the students. Flexibility was the key term students used and in today's world—where students must work longer economic hours to pay for university fees—it is a very practical idea.

There still remain some fundamental issues with the formats and the delivery of the lectures; many students expressed their concern over the lack of face-to-face teaching. Something that must be stressed here that this is not intended to replace that method of teaching only to support it; work is needed to get this balance right. On the whole, although the download times were slower, students preferred to use the higher resolution formats of the lectures.

Technology required:

- *Camtasia Studio*; *GarageBand*; Desktop microphone
- Webcam; Computer/s to make and for students to access; Online module space

Approximate cost of technology:

- *Camtasia Studio*—£230; *GarageBand*—£16; Webcam—£10–40 (at time of study)

What evidence is there of the effectiveness of this activity?
At the end of the course an anonymous questionnaire was placed on the module homepage on the Intranet service. An incentive was given to the students to complete the questionnaire by offering them a chance to win an MP3 player or a £20 iTunes voucher. 94 out of the 120 completed the survey (78 %).

Q. Most downloaded format:

- 86 % of students watched the high resolution screencast.
- Of those who watched the high resolution screencast—58 % watched it on the computer labs on campus compared to the 40 % who watched it off campus via a broadband connection.
- 20 % of students accessed the audio only format, of that 90 % downloaded the high resolution version.

Q. When they had accessed the online lectures during the course:

- 45 % accessed all or most of the lectures within the week of release of the new material.
- 34 % started to access them regularly and then failed to keep up.
- 21 % only accessed the online lectures the week before the exam.

The students were then asked open-ended questions and were given text boxes in which to write their answers. The main points to come out of this were; they liked the flexibility, and the dimension to pause and rewind the lecture (either to make notes or to go over something they did not understand). They enjoyed the freedom of not being held down to a lecture time and found it a vital revision tool. Some comments are as follows;

"I liked being able to pause and rewind bits so I could write down notes. I really struggle writing notes in lectures"

"I liked the idea of having it online so you could watch it when you wanted. Sometimes because of 9 o'clock starts you feel tired and don't take in as much."

Pitfalls/Problems/Limitations:
There were negative comments on the use of online lectures. A key point was that students often felt as if they were isolated and that it took them twice as long to go over lecture material than it would to attend the lecture because of the constant stop/start of the podcast material. Many students noted that unless they were highly motivated, it was easy to put it off and watch it another day and that they did not like not being able to ask questions there and then.

Key comments:
"I fell very behind with the lectures and it also took about double the time of the lecture to actually watch it as I became obsessed with catching virtually every word."

"I found I was writing every detail down which took forever, whereas in the real lecture you write key points you need to know."

Students also identified some problems, which they did not tell us when the lecture series was active. They complained that the files were often slow

> to download and some made note that the files did not work sometimes on their own media devices such as MP3 players.
>
> **Technological expertise required:**
> Staff skills required: Must have a basic level of IT skills, teaching of how to use techniques such as 'screencasting' and editing audio should take no longer than one hour.
>
> Student skills required: Little to none; as long as the student understands how to access the podcasts through the university intranet.
>
> **Also see:** Winterbottom (2007)

We also refer interested readers back to Case Study 3, from tutors at the Field Studies Council as that touched on the use of video.

5.3 Top Tips for Better Videos

A few simple 'top tips' will help make sure that there is reasonable video quality for tablets. It is a good idea for students to practise before they start on a project. Editing of video can be done subsequently with an app such as *Splice*.

- Don't wave the camera/tablet around, pan slowly, and limit the pan.
- Use a tripod; even a small one will help.
- Do not try to walk towards the subject in order to 'zoom'. Go closer, take some more video and splice them together.
- Poor audio is as bad as wobbly images. A small external microphone can be used to get good audio. Use a wind shield if possible. A lapel microphone (*lavalier*) with a clip can be brought to the speaker via an extension cable.
- Wind noise is very distracting. Adding audio later may be the best way to achieve good results.

5.4 Summary

This chapter provides a number of worked case studies that demonstrate the effective use of video, whether it is in the Scottish Highlands (Case Study 15) or the foreland of the Fox glacier in New Zealand, the result is to enhance the student learning experience. Good practice guidance with regards to capturing video in the field is provided by Case Study 13. This should ensure the best possible results. The potential of podcasting to support fieldwork learning is demonstrated through Case Studies 16 to 18.

References

Asset (2014) Moving forward through feedback: exploring the use of video to support and enhance the feedback experiences for staff and students. Retrieved from http://www.reading.ac.uk/videofeedback

Cake MA (2006) Deep dissection: motivating students beyond rote learning in vetinary anatomy. J Vetinary Med Educ 33(2):266–271

Cox B, Macharia R, Short N, Whittlestone K (2008) Podcasts and resources. In: Salmon G, Edirisingha P (eds) Podcasting for learning in universities. Open University Press and SRHE, London, pp 103–112

Foothgill J (2008) Podcasts and online learning. In: Salmon G, Edirisingha P (eds) Podcasting for learning in universities. Open University Press and SRHE, London, pp 80–91

France D, Wakefield K (2011) How to produce a digital story. J Geogr High Educ 35:617–623

Fuller IC, France D (2015) Securing learning using a 21st century Cook's tour fieldtrip. J Geogr High Educ 39: 158–172

Fuller IC, France D (2014) Fieldwork going digital. In: Thornbush M (ed) Geomorphological fieldwork, developments in earth surface processes, vol 18, Elsevier, pp 117–128

Fuller IC, France D. Does digital technology enhance student learning in field-based experiments and develop graduate attributes beyond the classroom? J Geogr High Educ (in press)

Gannod GC, Burge JE, Helmick MT (2008) Using the inverted classroom to teach software engineering. In: Proceedings of the 30th international conference on software engineering. ACM, Leipzig, Germany, pp 777–786

Hamersley B (2004) Audible revolution, the guardian. Retrieved from http://www.theguardian.com/media/2004/feb/12/broadcasting.digitalmedia

Heilesen SB (2010) What is academic efficacy in podcasting? Comput Educ 55(3):1063–1068

Impala (2012) User exemplars based on different approaches of using podcasts adopted by IMPALA 2 colleagues. Retrieved from http://www.impala.ac.uk/impala2/outputs/index.html

Jarvis C, Dickie J (2009) Acknowledging the 'forgotten' and the 'unknown': the role of video podcasts for supporting field-based learning. Planet 22:61–63. Retrieved from http://journals.heacademy.ac.uk/doi/full/10.11120/plan.2009.00220061

Jarvis C, Dickie J (2010) Podcasts in support of experiential field learning. J Geogr High Educ 34(2):173–186

Jenkins M, Lonsdale J (2008) Podcasts and students' storytelling. In: Salmon G, Edirisingha P (eds) Podcasting for learning in universities. Open University Press and SRHE, London, pp 113–120

Kemp J, Mellor A, Kotter R, Oosthoek JW (2012) Student-produced podcasts as an assessment tool: an example from geomorphology. J Geogr High Educ 36(1):117–130

Lee MJW, Tynon B (2008) Podcasts and distance learning. In: Salmon G, Edirisingha P (eds) Podcasting for learning in universities. Open University Press and SRHE, London, pp 92–110

Nie M (2008) Podcasts to support student learning in the GEES subjects. Planet 20:56–59. Retrieved from https://lra.le.ac.uk/handle/2381/4254

Salmon G, Edirisingha P (eds) (2008) Podcasting for learning in universities. Open University Press and SRHE, London

Traxler J (2009) The current state of mobile learning. In: Ally M. (ed) Mobile Learning: Transforming the delivery of education and training. AU Press, Athabaska University, Edmonton, pp 9–24

Winterbottom S (2007) Virtual lecturing: delivering lectures using screencasting and podcasting technology. Planet 18:6–8. Retrieved from http://journals.heacademy.ac.uk/doi/full/10.11120/plan.2007.00180006 http://journals.heacademy.ac.uk/doi/full/10.11120/plan.2007.00180006

Chapter 6
Social Networking, Communication and Student Partnerships

Abstract This chapter describes some of the ways in which mobile devices can be used to facilitate communication while on fieldwork. Social networks provide an informal way to share information by communicating within public networks (as in Case Study 19). Additionally, closed groups within social networks can provide a private online learning environment (as in Case Study 20) or to provide an environment for role-play or simulation exercises that require a communication channel (as in Case Study 21). Ways of communicating with and engaging an audience during fieldwork are also described. The final section is about student partnerships and how these can be used to develop novel learning tools.

Keywords Social networking · Student communication · Student partnership · App development · Online learning environment

6.1 Introduction

If there is one area of general smartphone use in the last few years it is in 'social networking. In particular, *Twitter* and *Facebook* are popular ways of communicating, especially the latter by school students. However, there are changing trends in popularity and it has been suggested that these are on the wane in popularity with other networks increasing in usage. There may have been a reluctance by tutors to use these media for student communication because these networks are seen as 'for students' (Roblyer et al. 2010). Fieldwork, however, is an area that would seem to be particularly suited to student-student-staff interplay, as well as for providing immediate comments and feedback.

© The Author(s) 2015
D. France et al., *Enhancing Fieldwork Learning Using Mobile Technologies*,
SpringerBriefs in Ecology, DOI 10.1007/978-3-319-20967-8_6

6.2 Social Networking

The three case studies outlined below describe the ways in which social networking has been used for student reflection and communicating findings on fieldwork. Case Study 19 reports on how information can be shared within public spaces within several Web 2.0 social networks. However, groups on fieldwork can also utilise private groups within the same social networks to access the same functionality and ease of use but out of the public domain. These can provide an online learning environment for reflective learning (Case Study 20) or provide an environment for role-play or simulation exercises (Case Study 21).

Case Study 19: Using Twitter, Qwiki and Flipboard in Student Fieldwork (Peter Bunting and Carina Fearnley)

Institution: Aberystwyth University, UK

Keywords: Mobile technology, fieldwork, teaching, learning, GIS, iPad, multimedia, *Twitter, Flipboard, Qwiki*.

Aim: To assess the usefulness of iPads and social media for field teaching within Geography.

Social networking through mobile devices is now widely accepted and used. This technological advancement can be used to develop fieldwork data collection. The rationale for this project stemmed from a field course that was led by the authors in New Zealand, in April 2011, but also other field courses undertaken within the department. It was felt that mobile technologies could have enhanced the fieldwork learning experience during these courses.

To evaluate how useful mobile technology and social networking would be in fieldwork, and if they would help to enhance the fieldwork experience, a project was developed which included a one day workshop undertaken with five undergraduate and postgraduate students. The students undertook similar exercises to those that they undertook within field courses they attended in their academic year. The workshop included an introductory session to discuss the aims and how the activities of the day fit in with their prior learning experiences on fieldtrips and excursions. The workshop involved the usage of iPads and social networking applications. The students were then asked to complete an online questionnaire to determine how their learning experience was affected.

Specifically, we wanted to answer the following questions:

- Can iPad's be used in place of paper based field handbooks?
- Can social networking be used to enhance learning and understanding in fieldwork, as well as enhance employability skills?

- Does the use of an iPad enhance the students learning experience?
- Does the use of an iPad encourage students to connect the theory with practice?

For social networking activities, an exercise was created to interact with a number of different applications including *Twitter*, *Flipboard* and *Qwiki*. The first stage of the activity was to use *Twitter* to gather and share information. An account titled @AberNZFieldtrip was set up, and the students were given the password to login to *Twitter*. They were asked to copy in relevant data and web links to information relating to Aberystwyth town centre. Once 20 min had been spent finding information and Tweeting the links, the students had time to review the information that had been posted by the others in the group.

Once the data were in *Twitter*, they used an application called *Flipboard* to see their data presented in a different format. Users were able to 'flip' through their feeds and those from websites that have partnered with *Flipboard*. Therefore a 'magazine' was set up in *Flipboard* of the tweets issued under @AberNZFieldtrip. *Flipboard* is designed specifically for the iPad, so the students used the iPads to view the books created. *Qwiki* was then used to look up areas of interest around Aberystwyth. *Qwiki* is a platform that creates interactive, on-the-fly, multimedia presentations of information. The students' experimented using different locations and names for places in Aberystwyth.

These activities were followed by discussion about the potential use of blogs or photoblogs, to reflect on the day's field activities. This is a virtual form of keeping a diary to reflect upon their experiences and thoughts.

Technology required:
A sample information pack was developed for the workshop. This pack included multimedia content that was developed for use on an iPad through a number of applications, including *Google Earth*, *Twitter* and *Flipboard*. Therefore computers are needed to generate the information and this is then used on the iPads by teachers and students in fieldwork.

Approximate cost of technology:
Computer: £200–£400.
iPad: £300–£500 (at the time of study).

What evidence is there of the effectiveness of this activity?
To assess the effectiveness of the project an online survey was undertaken with the students. The students felt that the iPads were a real advantage for fieldwork, particularly for displaying and manipulating mapping data compared to paper maps. The survey concluded that the iPad provides a useful tool that the students enjoy using, and it caters for multiple learning styles. It didn't take the students long to learn the user interface of the iPad and use the device correctly. iPads currently provide a very useful addition to the tools which can be used in teaching in the field, but many of the software applications available and reviewed in the project are only suitable for specific tasks or exercises, and cannot yet replace the paper field handbooks currently used.

Pitfalls/Problems/Limitations:
Students highlighted problems with sun glare on the screen of the iPad. The workshop was conducted on a particularly sunny day in Aberystwyth, which added to the problem. One of the iPads had a matt screen protector (Screen Devil) which helped to reduce this problem but did not completely solve it. In the future with new technologies it is expected that these issues will be resolved. For example the Amazon Kindle is still very readable in daylight because of its black and white screen, but it does not allow colour or the multimedia content to be displayed (although Kindle Fire does support colour and multimedia content).

Technological expertise required:
Staff skills required: Must be able to produce the multi-media information pack that is used with the iPad. Must also be able to use the iPad and the applications including *Google Earth, Twitter* and *Flipboard*.

Student skills required: Students must be able to use the iPads and the various applications that are to be used on it.

Also see: Fearnley and Bunting(2011)

Case Study 20: Encouraging the Professional Use of Social Media (Facebook) on Fieldwork (Rebecca Thomas, Alice Mauchline and Rob Jackson)

Institution: University of Reading, UK

Keywords: biosciences, iPads, *Facebook*, online learning environment, blog

Aim: To develop an online learning environment and to encourage the use of social media for reflective learning.

The University of Reading runs an undergraduate microbiology fieldtrip to Akureyri, Iceland (described further in Case Study 7). In 2014, the fieldwork research activities were again supported through the use of mobile technologies. In addition, the social media platform *Facebook* was used to engage students in their learning and to facilitate communication between the students and staff in the large multi-national team. A good, accessible communication platform was required as there were four institutions involved with staff and students from the UK, Iceland, Germany and Belgium.

Facebook was chosen as it is the largest social media network in the world and has been used previously to foster group communication in module teaching in Reading. Moreover, our experience of using *Yammer* as the main communication tool in the previous year suggested that students were less engaged with that system. Thus, a closed group was created within *Facebook* and all involved with the fieldtrip were invited to join. Initially it was used to help prepare the students for the trip; allowing the different student groups to communicate informally (this didn't happen to any large extent) and for staff to post relevant information.

Once in Iceland, our main aim was to encourage students to interact within the closed *Facebook* group as an informal, private, online, learning environment. The students were encouraged to post reflections on their learning experiences in the field (including photos and weblinks) and staff posted exemplars for them to copy. Regular 'breakfast quiz' questions were posted by the staff to encourage discussion online amongst the group. Also, a discussion seminar was hosted within the *Facebook* group; all students and the staff member were in the same room, but the students worked together in small groups with the *Facebook* group page projected on a screen at the front of the room. The staff member posted questions and the student groups replied with their answers, facilitating very active discussion which was captured in the online learning environment for future reference.

Technology required:

- Access to the internet (each student was provided with an iPad to allow easy access throughout the trip)
- A *Facebook* account

What evidence is there of the effectiveness of this activity?
During the course of the fieldtrip, all students and staff used the closed *Facebook* group to share or receive information. By the end of the two weeks, 93 % of the 34 students reported enjoying using the iPads to support their learning and 90 % felt comfortable using the *Facebook* group to communicate with others. The experience they gained using the closed group increased their confidence in posting about their learning on a social media site to a closed group of tutors, however there was no difference in their willingness to post on a public social media site. Nevertheless, at the end of the fieldtrip, the UK students were asked to write an assessed blog post about the benefits of fieldwork for microbiologists in a multi-national setting that was made publically available online. This exercise was designed to encourage the students to write about their experiences for a new audience and to provide an opportunity for them to start to develop their professional online identities.

Pitfalls/problems/limitations:
Some students may be reluctant to use *Facebook* or to use their own social account for their learning. This can be circumvented by pairing these students with others who would willingly share the information through other means and by explaining that they don't need to become *Facebook* 'friends' with staff to participate in the group. Also, providing each student with an iPad prevented anyone being disadvantaged if they hadn't brought along their own laptop, smartphone or tablet device.

Technological expertise required:
Staff skills required: Familiarity with the use of social media
   Student skills required: Familiarity with the use of social media and the basic use of iPads

Also see: Thomas et al. (2014)

Case Study 21: Using Web 2.0 (Yammer) Technology for Hazard Response Simulation Exercise (Servel Miller and Derek France)

Institution: University of Chester, UK

Keywords: real-time scenario, hazard management, *Yammer*, mixed methods, Web 2.0 technology

Aim: To develop a real-time simulation exercise using a closed group within a social media network.

Undergraduate students on the Natural Hazard Management programme at the University of Chester study volcanic hazard processes, management and responses in their final year of study. The students visit the Bay of Naples (7 days), Italy, in the first term of the academic year to gain an understanding of the spatial/temporal distribution of volcanic episodes and their underlying causative processes, their societal impacts, and the steps and processes involved in emergency planning. Students undertake an in-depth volcanic study of Mt Vesuvius and Solfatara. One of the challenges for planners in this region is how to effectively respond to and communicate volcanic hazard risk, should such a hazard become a reality in the future (the last volcanic eruption occurred over 60 years ago). A key part of the Chester programme of study is to introduce students to volcanic disaster response and effective risk communication. This case study provides an example of how students and staff on return to the University can utilise micro-blogging communication in hazard response simulation. *Yammer* is one of many Web 2.0 technologies that may be utilised in learning and teaching, such as role-play and simulation exercises. The benefits of this tool are that it is freely available and the interface is similar to most social networking sites used by students, such as *Facebook*. It was developed for corporate communication and, as such, operates as a closed system, limiting communications within specified groups (see Miller and France 2013).

Technology required:
Sign up to the *Yammer* Social network. Free access available linked to the same corporate email service account. A fee maybe required for Premium account users. Web access required and terminals, Desktop or Mobile devices can be used.

Approximate cost of technology:
Yammer is available free over the web.

Time was taken to develop this and implement the method:
Initially after conception and mapping activity, sign up a few hours i.e. quick. Students may need time to "play" with the system or have a mock exercise to familiarise.

What evidence is there of the effectiveness of this activity?
This has been comprehensively evaluated over a number of years from 2010 onwards when this intervention was introduced. The simulation exercise offered both interactivity between students and staff and, to a limited extent, between students and students. The preliminary analysis of the results from student feedback highlighted that they engaged well with the simulation exercise. The challenge of introducing a new type of assessment, which tries to facilitate a hazard simulation event in a controlled environment, was rewarding for both staff and students alike. While students found the experience stressful, they also believed it was exciting and helped them to better understand critical decision making during an emergency response situation.

Pitfalls/problems/limitations:
Valuable lessons have been learned regarding the simulation exercise process itself, preparation for the exercise, use of the exercise as an assessment, the (in)adequacy of computing facilities, students' interaction/lack of with each other and the communication tool itself (*Yammer*). The recommendations are:

Students should work in pairs or small groups, rather than as an individual, with one student consulting the resources pack, discussing their findings with their partner(s), and another student responding using *Yammer*.

The high level of 'web traffic' (over 100 micro-blog postings per session) meant a short delay in receiving messages consistently in a timely manner. The staff will consider decreasing the number of questions posed.

The duration of the exercise should be increased from 45 min to 1 h. *Yammer* should be introduced much earlier to students—at least at the start of the module. This will give them more time to use it to communicate with each other and become more familiar with *Yammer* as a communication tool.

Change the assessment of *Yammer* away from the quality of the blog postings to individual reflections on the process of using blog postings as reflective evidence. This will bring the assessment focus to the role of being a Hazard Analyst Officer and away from the operations of *Yammer*.

Technological expertise:
Staff skills required: Basic ability to type affectively and work from a prepared script will help
Student skills required: Information analysis skills, quick typing speed

> **Key tips or advice to others:**
> Create a working plan, develop a communication pathway and get your colleagues to participate in role play simulation exercise.
>
> **See Also:** Alistair et al. (2010), Miller and France (2012, 2013).

6.3 Communication

The following sections describe examples of how to communicate with and engage student groups in interactive ways. Face-to-face communication can be enhanced through audience participation using their devices. QR codes provide an opportunity for student interaction and a portal to access additional information either as text, via the internet or video resources. Word Clouds provide a way to quantify and display students' collective responses visually; with the larger fonts indicating the most frequent response.

6.3.1 Personal/Audience Response Devices

Personal response devices are a more sophisticated way of asking students to put up their hands in response to a question asked. For all sorts of reasons this questioning does not tend to elicit a good response. This is especially true for large classes and in the field. An alternative might be to ask students to show a coloured screen on their device, such as can be done with the *Flashlight* app. Or students could text or use *Twitter* with an appropriate hashtag. Once again ingenuity and finding out what colleagues have used is one way to exploit the technology.

In recent years, electronic devices (PRS or 'clickers') have gained a following to elicit not only responses but answering multiple choice questions. However, these are costly, difficult to set up and although if your institution has a set, there are better ways using smartphones and iPad apps. One such is *Polleverywhere*; this app collects information that comes from mobile devices, *Twitter* or web browsers. The app can be run free for up to 40 users but it may be very economic for an institutional or departmental subscription. *Socrative* and *eClicker* offer similar functionality. *Polldaddy* allows you to conduct surveys of some sophistication on a mobile device. Another application to consider would be *Everyslide* which facilitates audience participation through their mobile devices.

6.3.2 QR Codes

QR, Quick Response, codes have been commonplace in publications, advertisements and even conference badges in the last five years. As a somewhat more sophisticated form of bar code they allow a mobile, internet-connected, device to go straight to a web-located source, whether that be website, video or an e-mail address. Apps to read the code (and usually bar codes additionally) are usually free and easy to use. Even if you are not internet-connected, a photograph can be taken and used later. The code readers operate over a surprisingly wide range of sizes so that it is possible to make the link by photographing a projected QR image or on a monitor or TV screen. The significant aspect of QR codes is that they can be generated very rapidly by using a free website (such as http://www.qrdroid.com/generate). Generating a code (via an internet connection) produces an image that can be used in all sorts of ways.

A quick web search will provide several imaginative teaching applications for QR codes. For example they might be used as an icebreaker treasure hunt, around a field centre for tree identification or displaying local site information. A benefit of using QR codes on (semi)permanent signs/posters is that these can be kept up-to-date by keeping the on-line information updated. As well as using QR codes to access materials, students could be creative in generating QR codes for specific uses in their projects.

6.3.3 Word Clouds

Word clouds, sometimes known as a *Wordles*, as in the example below, are frequently used to highlight words commonly used in a piece of text. Other, similar apps are *Cloudart* and *Wordsalad*. Questionnaire responses are a useful way of summarising and comparing data sets that might have been gathered on a fieldwork exercise. In effect, word clouds provide a semi-quantitative measure of textural analysis.

Word cloud apps can deal with the words and phrases in slightly different ways and present their results in various configurations (Fig. 6.1). Getting students to explore their use provides a good opportunity for experimentation in information presentation and visual design. For example, colour blindness is a surprisingly common complaint and that red is by no means a good colour for visual presentations in general.

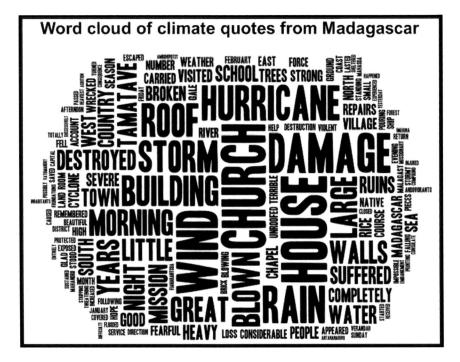

Fig. 6.1 Word Cloud from qualitative data from missionary diaries in Madagascar 1865–1900. (Courtesy of Prof. David Nash, University of Brighton)

6.4 Student Participation

The two case studies in this section describe how student partnerships can be used successfully to develop tools (apps) that can support student learning. There is currently a drive for students in Higher Education to become co-creators of knowledge and to become active participants in curriculum design (Bovill and Bulley 2011). Both case studies achieve this by drawing on students from different faculties to work in multi-disciplinary teams on mobile app development projects. As well as creating the learning tools, the students involved gained many skills that are relevant for their future employment including communication skills, project and time management skills.

Case Study 22: SpeciesHack 2013—a Multidisciplinary Hackathon to Develop Apps for Biology Fieldwork (Judith E. Lock, Moira A. Maclean, Chris D. Sturdy, Davide Zilli and Alex C. Rogers)

Institution: University of Southampton, UK

Keywords: smartphone, tablet, mobile applications, hackathon, species identification

6.4 Student Participation

Aim: To investigate the use of student partnerships from different faculties to design and develop apps that could be used on a field course. This student-focused, interactive project involved two app-creation events coined as 'hackathons'.

The Centre for Biological Sciences' first year field course, focusing on Experimental and Field Ecology, takes place in Spain. The amount of equipment that can be taken is limited, so the course has traditionally been very lo-tech with presentations taking place using paper table cloths. Weight restrictions on the flight to Spain also limit the number of field identification books that can be taken.

Smartphones and tablets (which many students already own) provide a new opportunity for beneficial use of technology in the field, allowing data collection and analysis (including species identification) in locations that were not previously possible. To some degree, the apps presently available for the smartphones and tablets do not provided species identification guides that are easy to use and comprehensive enough for accurate identification. This is where a student-focused hackathon can both identify and attempt to provide essential apps that are not already on the market; potentially providing tools for fieldwork use that can be distributed commercially and financially support future projects.

A student hackathon requires students from two cohorts: (1) those that provide the subject-orientated ideas, in this case those who had been on field courses, the 'idea' students; (2) technically-minded students who can turn the idea into an app, the 'developer' students. Our first cohort came from the Centre for Biological Sciences and Ocean and Earth Sciences. Our second cohort came from Electronics and Computer Science.

The date and aims of the hackathon were advertised to both groups of students with a link to a website where they could register for a limited number of spaces on a first come, first served basis until we had 12 students from each group and a waiting list.

We began the event at 10 am with a mixer, encouraging students to interact and form their teams of four (2 "idea" and 2 "developer" students). The teams were told that prizes (£200 of Amazon vouchers) would be awarded for three categories:

- Best prototype app
- Best designed app
- Best useable app

A deadline for app presentations to the other teams was set for 6:30 pm, followed by awards for prizes.

Technology required:
Each team was provided with pens, paper and A3 card for storyboarding their app.

Students were told to bring their own laptops (set-up with appropriate app-development software such as Android SDK and wireframing tool for the 'developer' students), smartphones and tablets. Those who had been on a field course (the 'idea' students) were asked to bring their field course notebooks.

Approximate cost of technology:
There was no cost for technology, as we followed the "bring your own device" route. However, with a longer event, app development could reach a further stage where testing using tablets would be beneficial.

Costs were associated with refreshments (we provided lunch and dinner—students commented that breakfast and ice-cream would also have enhanced the event!) totalling £500, and prizes totalling £600.

How much time was taken to develop this and implement the method?
About an hour a week throughout semester 1 (12 weeks), increasing to 6 h in the week before the event. We held the hackathon on a Saturday, which we all attended.

What evidence is there of the effectiveness of this activity?
Student comments showed that they really enjoyed interacting with students on different degree programmes. The 'developer' students particularly commented that it was nice to develop an app with a purpose. The 'idea' students liked the opportunity to bring their ideas to life.

Pitfalls/Problems/Limitations:
Students should be aware of the limitations of a 1 day event, with little time to see the project come to fruition. After the initial idea creation stage there is a limited amount that the 'idea' students can do while the 'developer' students are coding. If we expanded to a 2 day event, we would have to think of an additional task for 'idea' students.

Technological expertise required:
Staff skills required: App development or field course teaching
Student skills required: App development or field course attendance

Key tips or advice to others:
The promise of free food is very motivational to students! We have not yet been able to use the apps on a field course, but we hope to attract an Electronics and Computer Science project student to undertake further development of the winning apps.

This next case study builds on the idea of student partnerships to develop learning tools that were generated in Case Study 22. In Case Study 23 the project funds were spent on providing student bursaries for which each student was expected to contribute 50 h to the project over 8 months. This longer timeframe enabled much more development time to produce a final, workable end product.

Case Study 23: The Development of a Campus Biodiversity Recording App: A Staff-Student Partnership (Alice Mauchline, Liam Basford, Stephen Birch, Alison Black, Alastair Culham, Hazel McGoff, Karsten Lundqvist, Philippa Oppenheimer, Jon Tanner, Mark Wells and Liz White)

Institution: University of Reading, UK

Keywords: multi-disciplinary, engagement, citizen science, app development

Aim: To develop and implement the use of a biodiversity recording mobile app in teaching activities at the University.

This project involved a multi-disciplinary group of 6 students working in partnership with 5 staff members all from different Faculties. It was funded by the Teaching and Learning Development Fund at the University of Reading and the project provided a way for students to develop new tools to enhance their learning experience.

Species identification and biodiversity monitoring techniques are taught on the University campus in a range of modules across several schools. To support and integrate these activities, the Whiteknights Biodiversity Blog was established in 2011 as a central repository for information about campus biodiversity monitoring, however the data weren't collected in a central database. Therefore the staff-student team worked together to develop a mobile app for recording geo-tagged field records on campus. The project was based on a citizen science approach in order to foster the on-going staff-student partnership as it allows easy contribution from many individuals and results in a database that can be used to support teaching and research projects in the future.

The target end-users of the app were students and staff from Environmental Science and Biology and Agriculture, therefore they were responsible for the app content. The Typography and Graphic Communication student was supervised in creating the branding and improved the usability of the final app. Two students from Systems Engineering were supervised in developing the code to customise the app.

A scoping study was conducted to compile a list of basic requirements for the app to meet the need of target users. The team held a 'HackDay' to discuss the way in which the app could be developed and how the project should progress. We discovered that the basic functionality required for this app is available in the *EpiCollect* app (see Case Studies 9 and 10) and importantly the underlying code is open for developers to modify. *EpiCollect* was particularly suitable as it works offline to record data, has inbuilt GPS location recording and has the ability to sync records over Wi-Fi to our server.

An important element of the project was to bring individuality and branding to the app to allow easy identification of the product by a community of users. Branding was also important in the creation of a supporting website (KiteSite 2014) which offers information and user support to augment the app. The website also provides a hub for information such as later developments, aided by the implementation of a *Twitter* feed, which provides a newsfeed and creates a community discussion point. The website also houses links to the data and maps, allowing access from any platform. The prototype app was subjected to several levels of user testing and refined before the final product was launched to staff at the University.

Technology required:
App development: software for writing iOS (e.g. xCode) and Android code (e.g. Android SDK).

Use of the mobile app: any iOS or Android mobile device (internet connection not required for data collection in the field)

What evidence is there of the effectiveness of this activity?
At the launch workshop, several staff members discussed ways in which they anticipate integrating it into their teaching activities and it is already being used in MSc teaching and to support environmental science student projects. Also, student natural history groups have shown an interest in using the app to record sightings on campus outings.

Pitfalls/problems/limitations:
There was some difficulty customising the iOS version of the app and there were other limitations that prevented embedding of the data map in forked versions (versions generated from the *EpiCollect* source by third party developers). However, designing an app from scratch may be the only way to fully customise such software.

Technological expertise required:
Expertise in writing software code for iOS and Android systems is required for app development. A basic knowledge of mobile devices is required to load and use the app.

Also see: Mauchline (2014), Whiteknights Biodiversity Blog (2014) and White et al. (2015)

6.5 Summary

This chapter has presented various ways in which technology can be used to enhance communication during fieldwork activities and to foster staff-student partnerships. Social media will continue to be an important way for staff and students to interact and fieldwork provides a unique atmosphere to encourage informal dialogue. Strong communication skills are important for graduate employability and fieldwork activities can provide a good group size in which students can strengthen and develop such attributes.

References

Alistair M, Gemmell D, Finlayson I, Marston P (2010) First steps towards an interactive real-time hazard management simulation. J Geogr Higher Edu 34:39–51

Bovill C, Bulley CJ (2011) A model of active student participation in curriculum design: exploring desirability and possibility. In: Rust C (ed) Improving student learning (18) global theories and local practices: institutional, disciplinary and cultural variations. Oxford Brookes University, Oxford, pp 176–188. ISBN 978-1-873576-80-9

Fearnley C, Bunting P (2011) Understanding the potential new roles for mobile computer technologies for teaching geography fieldwork: using GIS and social networking to enhance learning opportunities. Enhancing fieldwork learning showcase event. Port Talbot, Wales. pp 16–18 September 2011. Retrieved from http://nexus.aber.ac.uk/xwiki/bin/view/Main/Enhancing+fieldwork+learning

KiteSite (2014) Retrieved from www.reading.ac.uk/herbarium/kitesite

Mauchline AL (2014) Retrieved from https://blogs.reading.ac.uk/engage-in-teaching-and-learning/2014/10/23/are-you-interested-in-biological-recording-monitoring-with-your-students-by-dr-alice-mauchline/

Miller S, France D (2012) Developing a Web 2.0 technology for hazard response simulation, higher education academy STEM annual conference paper. Retrieved from http://journals.heacademy.ac.uk/doi/pdf/10.11120/stem.hea.2012.045

Miller S, France D (2013) Real-time emergency response scenario using a Web 2.0 technology. Planet. 27(2):21–29 HEA GEES. Retrieved from http://journals.heacademy.ac.uk/doi/full/10.11120/plan.2013.00005

Roblyer M, McDaniel M, Webb M, Herman J, Witty JV (2010) Findings on facebook in higher education: a comparison of college faculty and student uses and perceptions of social networking sites. Int Higher Edu 13(3):134–140

Thomas RL, Mauchline AL, Jackson RW (2014) Facebook, iPads and 'extreme' microbes in Iceland. Retrieved from http://blogs.reading.ac.uk/engage-in-teaching-and-learning/2014/08/27/facebook-ipads-and-extreme-microbes-in-iceland-by-dr-becky-thomas-dr-alice-mauchline-and-dr-rob-jackson/

White, E, Basford, L, Birch, S, Black, A, Culham, A, McGoff, H, Lundqvist, K, Oppenheimer, P, Tanner, J, Wells, M & Mauchline, A (2015) Creating and implementing a biodiversity recording app for teaching and research in environmental studies. J Educ Innovation Partnership Change 1(1). ISSN 2055-4990

Whiteknights biodiversity blog (2014) Retrieved from http://blogs.reading.ac.uk/whiteknightsbiodiversity/

Chapter 7
Pre-field Trips and Virtual Field Trips

Abstract In this chapter we consider the potential of virtual field trips/guides, reusable learning objects and augmented reality to support fieldwork learning. A variety of case studies are presented to show the necessary development time, support and skills required as well as evaluations of their use by students.

Keywords Reusable learning objects · Virtual field trip · Augmented reality · E-learning · Field guide

7.1 Introduction

As well as 'actual' fieldtrips of many kinds still being an important part of many subjects in Higher Education, it is also important to note that the Web has introduced the idea of 'virtual' trips. This possibility was noticed over ten years ago. Some early examples (which still exist) of Virtual Field Trips (VFT) were images and text on web pages. While producing such a guide may be perfectly fine for some purposes; it is not 'virtual' in the sense of immersion such as 'second life'. The simple pages are perhaps best called 'Virtual Field Guides' (VFG) for that is what they are. Originally they could be viewed on web pages and perhaps downloaded as PDFs. Now, with tablet technology they can be similarly downloaded but perhaps as e-books rather than PDFs. For this they are undoubtedly useful, especially in the planning stage, to show students what might be encountered or where to go on a site for further investigation. Such guides could be linked in with geo-referencing or geocaching techniques so students could find their own way to locations and investigate as appropriate. This would provide a good opportunity for self-guided trips that students might need to make. Furthermore students could prepare their own survey of an area, to acquaint themselves to it, before departure. In essence, this is another form of the poster preparation mentioned in Case Study 4 in Chap. 3.

However, the imagination and enthusiasm for VFTs somewhat ran aground on the lack of computer memory and processor speed. This limitation was especially noticeable when large classes were involved. VFTs have always been intensive of preparation time and, because they were generally site specific, not very adaptable. However, some remain available and examples are given below. Some are web-based resources that lic between a static/video guide and an excursion. An example is the 'Virtual eruptions on a supercomputer' (Evivo 2014) where compilations from computer runs have been made available. Nasa has a variety of such tours; an example is the Curiosity Rover (2014). This uses the *Photosynth* app for making 3-D composites.

Some VFTs, such at the Open University's geological trip around Skiddaw (see Virtual Skiddaw 2014) in the English Lake District are very feature rich—looking at rock outcrops in this case with a variety of tools. They allow fly-by techniques to provide an overview or recapitulation of a project.

The tools provided by *Google Earth* for constructing virtual trips as well as other forms of visualisation are perhaps the most useful way for tutors and students to construct their own trips. A useful compendium for geoscience education are the contributions to the book by Whitmeyer et al. (2012). There is no doubt that students' creative ability, when coupled with digital and information skills, can be enhanced by creating their own tours. The result might report on what they did to produce a guidebook and enhance their employability skills as a consequence. Some of the technical aspects, such as creating KML files, can be done fairly easily and details are given in some of the chapters in Whitmeyer et al. (2012).

Chapter 8 contains some virtual approaches to fieldwork, especially where disabled students are concerned. But actual access to a site may not be practicable for anybody. To this end, *GigaPan* technology (think of a zoomable *Google Earth* view taken from the ground) can be used and described in more detail in Chap. 8. A paper by Stimpson et al. (2010) provides a good example. *GigaPan* views are, at present, only available on the *GigaPan* website so outcrops can only be viewed before or post trip. Examples are the igneous intrusions in the vertical Scrabo sandstone cliffs Whalley (2010), Stimpson (2009), which shows brecciated rhyolite at a close up but on a rather inaccessible coastal site.

7.2 Re-usable Learning Objects and Fieldwork

A re-usable learning object (RLO) is closely related to Open Educational Resources (OER) and the terms are sometimes used synonymously. There is a good discussion about terms and usage by Polsani (2003). In brief, a learning (or educational) object might be an image you take. You can then use it in several presentations or posters or whatever. It might be repurposed, perhaps with annotations added. It becomes 'open' if you allow its release according to a (usually) Creative Commons (CC) licence. The image will usually encompass the associated metadata; date, time, author, subject etc. By making it an OER it allows others, with accreditation, to use or to repurpose it.

In the UK, the JISC JORUM (see Jourm 2014) provides access to a wide range of OER resources. Of course, the resource can be much more than an image; a video or several slide presentation (and see also *Slideshare*) or, as below, a field guide or trip.

This next case study, by Phil Porter, shows how students can contribute to communal websites that can be developed. This is a good example of student involvement using digital technologies. It should be noted that this project was done before the days of iPads and relatively expensive technology was used for the project. We have mentioned ways in which iPad technologies can capture good quality video and edit it in Chap. 5.

Case Study 24: Linking Teaching and Field Research: Student Engagement with Reusable Learning Objects (Phil Porter)

Institution: University of Hertfordshire, UK

Keywords: Reusable Learning Objects, Fieldwork, Research-informed Teaching, Video

Aim: To engage students in the production of reusable learning objects, which demonstrate a variety of field techniques relevant to fieldwork research.

It is clear that fieldwork has a vital role to play within the disciplines of geography and environmental science and that the educational benefits are significant for both students and staff. Also of great significance is the unparalleled opportunity that fieldwork presents to engage students with research and to assist in furthering the integration of teaching and research (the so-called 'teaching-research nexus'), which also has clear mutual benefits for academic staff and students.

Having taught field-based Physical Geography at University level since 1996, a worrying trend observed has been an increasing reticence for students to undertake field-based research in pursuit of final year dissertation work. Informal feedback from students reveals that a lack of knowledge of field techniques that may fall outside those commonly taught in undergraduate programmes is perceived as a significant barrier, preventing greater participation in field-based data collection for subsequent use in the production of dissertations.

In an attempt to break down this barrier and to further the integration of teaching and research a project was set up to engage students in production of a suite of reusable learning objects, comprising a suite of videos with audio commentary that demonstrate, in step-by-step format, how to undertake a variety of techniques relevant to research fieldwork. The videos focus upon techniques that are not always commonly taught in detail at undergraduate level, but have strong relevance to potential student dissertation work.

The videos and associated content are hosted on a public website Fieldwork (2014a) which has been designed and built entirely by undergraduate students at the University of Hertfordshire.

The key components of the website are as follows:

1. A series of step-by-step 'how to do it' videos that take the viewer through a variety of field techniques in detail and are embellished where appropriate with commentary, diagrams and animations.
2. A series of 'top tips' videos, which provide experience-based tips for students to consider when undertaking fieldwork.
3. A gallery of inspiring images from the field, all taken by students.

Furthermore, with each video comes a brief text based explanation of essential equipment required, suggestions for possible student dissertation projects and a selection of useful links (e.g. websites and literature).

Production of these learning objects by students has multiple benefits, several of which link to established principles of good practice in undergraduate education (Chickering and Gamson 1987). Firstly, students work closely with academic staff when filming field techniques being practiced and some of these techniques are closely linked to staff research projects (encourages contact between students and faculty). Secondly, students work together in order to film and produce video footage (reciprocity and cooperation amongst students). Thirdly, students know that other students and academic staff will use their finished work, potentially worldwide (communicates high expectations). Finally, and perhaps most significantly, one of the driving principles behind the project is the belief that if students can 'see' their peers demonstrating and undertaking field research, this will engage students with field research and associated techniques and help encourage more students to 'have a go' and increase the uptake of field research based dissertations.

What technology is required?
Use of the website requires Internet access and a computer capable of playing *YouTube* files. Camcorders were used to film field footage and a semi-professional editing software package was used to produce video clips. High quality camcorders were used in order to ensure high-resolution output and avoid the significant limitations imposed by devices such as Flip Cameras and mobile phone based video cameras. Semi-professional editing software was used in order that additional content (e.g. animations, imagery etc.) could be included in edits without the limitations present in entry-level software.

Approximate cost of technology:
Canon Legria High Definition (HD) camcorders were used for filming field footage, at a cost of approximately £250 each (at the time of the study). Editing software (*Cyberlink PowerDirector*) cost approximately £100 for three licences. [Note that much lower cost alternatives for iPads now exist]

How much time was taken to develop this and implement the method?
Filming took place over a two-week period. Editing and production of clips took two students approximately two weeks of work (based on a 37 h week)

and website design and construction took a similar period, again with two students working on a full time basis but with an additional student assisting with graphic design aspects (e.g. animations for inclusion in video clips) for one week.

What evidence is there of the effectiveness of this activity?
Informal student feedback indicated that the website had achieved one of its key objectives, namely making fieldwork and specialist research techniques accessible to an undergraduate audience and providing motivation for the uptake of field based research. Although direct cause and effect cannot be proven, it is gratifying to have seen a higher percentage of University of Hertfordshire Geography and Environmental Management undergraduates undertaking field based research in pursuit of final year dissertations since the website was launched.

Users of the website area are able to contact the academic managers of the project; feedback gathered in this way has been unanimously positive. Comments left on our *YouTube* channel have also been positive in nature.

Pitfalls/Problems/Limitations:
Editing video footage is a time consuming exercise and sets the limit on what can be achieved with available funding and student time; filming is relatively rapid by comparison. This project is therefore viewed as an ongoing database of reusable learning objects that will be updated as and when students can be paid to work on video editing.

Technological expertise required:
Students were instructed in the basic operation of video cameras and self-learned editing skills through use of *Cyberlink PowerDirector* which was the editing package chosen, largely due to its user friendly interface, combined with powerful editing capabilities.

Students and staff who make use of the reusable learning resources need no specific expertise, although additional training in the use of field equipment shown on videos may be required.

Key tips:
Using a high quality camcorder and editing package is a must. The limitations of mobile phone video and devices such as Flip Cameras soon become apparent when filming and editing. Similarly, free online editing packages are likely to frustrate attempts to produce professional looking edits. *Cyber Link Power Director* (used on this project) is one of several editing packages that are reasonably economical to purchase, yet provides powerful editing capabilities. Industry standard packages (e.g. *Final Cut Pro*, *Adobe Premiere* etc.) could also be used and, although these may cost more and potentially take longer for students to learn, they have the advantage of providing students with experience in using software that they may come across in their employment.

> **Helpful documents and references:**
> A draft website, designed and built by students and showcasing the resources produced by them to date, can be found at Fieldwork (2014b).
>
> This website is not completed and is regarded as an ongoing resource that will be updated as and when students are able to upload edited video material when funding permits.
>
> **Also see:** Chickering and Gamson (1987)

7.3 Augmented Reality

Augmented reality (AR) provides a 'live' view of a physical object or view that is augmented or enhanced by the tablet or smartphone. Virtual Reality (VR) replaces the real world with a simulated one. AR enhancement might be typically audio (a piece of music for example or birdsong), video or still imagery. As might be expected, this is an area of considerable research (including *Google Glass*) not least in military applications. However, there are many applications (as downloadable apps for specific sites) in tourism (see Sung 2011). An innovative way of viewing the London Underground from above ground is provided by *NearestTube* (2014). AR is also used widely in advertising publishing and again a web search will provide many educational ideas. It is here that we can move to downloading video as a piggyback on a still image. You might have constructed an instructional video. This could be seen on demand by students, as they needed. Thus, an addition to the basic, student produced, poster (as in a previous case study) would be to use an augmented reality approach.

There are several apps that can be used to produce AR products. These include *Wikitude*, for which a software development kit (SDK) needed, *Layar* and *Aurasma* and *Junaio*. For the latter you use the *Aurasma* app on your smartphone. You will need a username and password but the website provides a tutorial on how to use it. This is an app that requires a WiFi link from a mobile device to a web source. In effect, this is image recognition just like a QR code but allows a picture to be created with embedded text, video etc. that is recognised by the device's camera.

The following case studies show different approaches to Virtual Field Trips and Guides and are specific to institutions. They also show the need for considerable input by tutors to set them up. However, once established they do provide students with surrogate field experiences, especially using *Google Earth*. It remains to be seen how the latest mobile technology, especially using collaborative work and cloud-based data, can be incorporated to enhance students' fieldwork experiences.

Case Study 25 Google Earth, Virtual Fieldwork and Quantitative Methods in Physical Geography (Varyl R. Thorndycraft, Don Thompson and Emily Tomlinson)

Institution: Royal Holloway, University of London, UK

Keywords: *Google Earth*, virtual fieldwork, quantitative methods, geomorphology, teaching, learning

Aim: To combine virtual fieldwork (using *Google Earth*) with the application of equations in spreadsheets, to improve the students' understanding of quantitative methods.

This case study discusses a module that incorporates both virtual fieldwork using *Google Earth* with the application of equations in *Microsoft Excel*. Two seminars were designed for the second-year undergraduate Geomorphology course and took place in a computer teaching laboratory with small students groups of below 20; incorporating both virtual fieldwork using *Google Earth* with the application of equations in *Microsoft Excel*.

One seminar, which was designed for the fluvial geomorphology section, focused on the concept of stream power; it did this by comparing how stream power changed downstream along the Brahmaputra River. The students had to calculate the stream power; to do this they had to determine the slope and flow discharge.

The data for the average flow were derived from a published journal article as the data were not available on *Google Earth*. Using *Google Earth's* measuring tool, the students were asked to calculate the slope of the river at each of the locations by measuring the longitudinal distance and spot heights. The main objective of this exercise was to engage the students with the equations and calculations.

Another seminar that was designed for the glacial geomorphology section focused on the dynamics and landforms of glaciers. This seminar used the shear stress equation. The students were provided with realistic hypothetical ice thicknesses for a variety of glaciers.

To calculate the shear stress, the students had to determine the slope angle of the glaciers using *Google Earth*. The students then compared the shear stress results from different parts of the glacier and related the results to glacial morphology. The students were also required to determine whether the glacier was advancing or retreating using their geomorphological evidence.

The application of quantitative methods in physical geography is an important skill for students, as it enhances their understanding of the subject as well as developing their generic skills including numeracy. Therefore new teaching and learning methods focused on the understanding of quantitative methods are becoming increasingly important.

The development of *Google Earth* provides an exciting new teaching resource for geography. It is already widely used and has become a key tool in many subjects, not only geography. *Google Earth* allows new opportunities for innovative teaching and learning methods focusing on virtual fieldwork.

Technology required:
The technology used in this method included *Google Earth* and *Microsoft Excel*, the other data provided to the students was acquired from other sources such as journal articles.

How much time was taken to develop this and implement the method?
The teaching method was developed and used in two seminars in computer laboratories. The technology used was already available. The method just had to be organised and prepared.

What evidence is there of the effectiveness of this activity?
To assess the two seminars, a student focus group was used. The group revealed that overall they were very positive about the incorporation of *Google Earth* in the classroom. The main positive reflection was that the students felt that the interactive seminars enabled them to engage with the landscapes and environments more than they would in a normal lecture. Students also said that *Google Earth* was a very successful teaching tool and that they would use it in the future, at home as well as in education.

The seminars were very successful in interesting the students and engaging them with quantitative methods, which improved their skills as well as their confidence.

It was outlined by the tutors that this teaching method of virtual fieldwork is not designed to replace fieldwork, but to act as a link between field and class work.

Pitfalls/Problems/Limitations:
One of the limitations of *Google Earth* is that the accuracy is not sufficiently high to enable precise height measurements. To solve this, the slope data along the Brahmaputra (first seminar) was provided for the students. The students could then compare their field data with the slope data provided for them. The aim of the exercise was to engage the students with the stream power equation.

This teaching method does not focus on the accuracy of the results, as its main objective is to engage the students with the quantitative methods including calculations and equations; but the tutors should still be aware of the inaccuracies of *Google Earth*.

Technological expertise required:
Staff skills required: Basic computer skills required. Must be able to use both *Google Earth* and *Microsoft Excel*.

Student skills required: Basic computer skills required. Must be able to use both *Google Earth* and *Microsoft Excel*.

Also see: Thorndycraft et al. (2009)

Case Study 26: Design, Development and Student Evaluation of a Virtual Alps Field Guide (Tim Stott, Anne-Marie Nuttall and Jim McCloskey)

Institution: Liverpool John Moores University

Keywords: E-learning, fieldwork, virtual environments, field Guide

Aim: To provide students with a Virtual Field Guide (VFG) to view supporting material for the fieldwork, including short instructional videos, images of the field site and *PowerPoint* presentations.

As with many Geography courses, fieldwork is an integral part of the learning process. It builds vital skills for students to take with them throughout the duration of their courses and provides vital skills for employability. With this in mind a VFG was created for the Alps. A basic but professional web template was used, ensuring that the lecturers would not be overwhelmed if they had limited web based skills. Upon completion, the virtual field guide would eventually be placed on the module-shared space for students to access. However first the VFG was evaluated with two groups of students:

- Second-year BSc Physical Geography/Geology students studying Glacial and Fluvial Processes in the Faculty of Science (n = 20)
- Third-year BSc Outdoor & Environmental Education students studying Glacial and Fluvial Processes in the Faculty of Education, Community & Leisure (n = 12)

The students volunteered to follow the instructional sheets guiding them through the field guide, which explained how to navigate its interface, before they were set tasks to complete within a set timeframe. Upon completion of each task the students were asked to provide anonymous feedback via Likert Scale based questions and write free text comments on each of the three virtual sites. Overall the students had an evaluation time of 1 h to complete the tasks and feedback sheets. Then the results were collated for evaluation.

Technology required:
A computer with *Adobe Dreamweaver* web-editing technology is required. This software was vital for allowing the lecturers to place the field guide in the correct format onto the Internet. Furthermore the *Dreamweaver* software allowed the lecturers to use the power of the software to easily edit the template and give it a professional look. *Dreamweaver* allowed the lecturers to create tables, insert imagery, video clips/panorama movies and create hyperlinks.

Other software such as *PowerPoint* and *Microsoft Picture Manager* is also needed.

RedDot software was also used to then upload the finished *Dreamweaver* file onto the university intranet. This software has been widely used for many years at the university.

Approximate cost of technology:

- Computer—£400–800
- *Adobe Dreamweaver*—£313 (educational discounts are available at present time)
- *RedDot* Software—Individual quotes and enquires only

How much time was taken to develop this and implement the method?
50 h was required for completion of the basic design and to upload the design onto the intranet. However continual maintenance and corrections are needed. However, maintenance is now the responsibility of a university IT technician.

What evidence is there of the effectiveness of this activity?
On the whole the students were very positive in their remarks about the virtual field guide. All students 'agreed' that the virtual field guide was user friendly and easy to navigate around and all 'agreed' or 'strongly agreed' that the pictures were of good quality.

The students also expressed that the exercises provided were important in helping them to develop their independent research skills. Most students also agreed or strongly agreed that further reading sections would have been useful and that they enjoyed the short revision quizzes that were at the end of the field guide.

Pitfalls/Problems/Limitations:
Once completed, the field guide was continually updated and maintained by one of the IT technicians in the university who carried out our work in his free time. However he left for another post making it increasingly difficult to find someone who would carry out his previous duties. This led to a stagnation of development on the field guide and, when a replacement was found, there was a long lag time between designing and asking for changes to it being implemented. The technician retains control and academics can be frustrated because they are not able to make simple changes or updates to the VFG.

Academic staff do not usually have the web skills or importantly the time to learn advanced webskills, leading to the field guide looking somewhat amateurish. Academics carry out administration, teaching, fieldwork and research and find it very difficult to prioritise the time to spend learning the skills and developing the VFG.

Technological expertise required:
Staff skills required: medium level web skills, importantly staff needed to be trained on how to use the complex *Dreamweaver* software. Also a highly trained IT technician is required to oversee the project.

Student skills required: Little to none.

See also: Stott et al. (2009), Stott et al. (2014)

Case Study 27 Learning Desert Geomorphology Virtually Versus in the Field (Richard J. Stumpf II, John Douglass and Ronald I. Dorn)

Institution: Arizona State University, USA

Keywords: Education, fieldwork, geomorphology, physical geography, virtual

Aim: To allow physically disabled students to learn about the landforms of Arizona and the Southwest, USA.

The field site used for this teaching method is Tempe Butte, known as 'A Mountain' because of the large 'A' built on the southern slope facing the campus. The field trip to Tempe Butte represents one of the Arizona Board of Regents (ABOR) modules and is specifically designed for students in the Introduction to Physical Geography course. A number of programmes use the site for student field experiences due to its close proximity to the campus, the site is ideal for the Desert Geomorphology Unit because of its desert environment within an urban setting. Tempe Butte is also used because of the differences between its north and south landforms, hydrology, biogeography, soils and geology.

Students who participated in the VFT were taken to a computer lab during their laboratory period. The teachers instructed them on how to access and navigate the virtual experience; they also provided additional support and clarification for the students if they required it. Students can access the virtual field trip through the Internet; it is presented as a series of web pages with a three-tiered structure, including the index, field trip stops and concepts.

The students are provided with a brief history of Tempe Butte on the index page, as well as a navigational structure created using an aerial photograph of the site, which shows the location, and view of each stop on the physical field trip. The students can click on each location through the aerial photo; they are then directed to a page, which includes the key concepts that are relevant to that location.

The panoramic photographs provide the students with the same view that students would have on the physical field trip; this picture includes text and coloured boxes indicating the key concepts. The students can click on the text to be directed to another page, which discusses the subject in greater detail, including close-up pictures, video footage and graphical models showing the geomorphic features in further detail.

Technology required:
A large number of educational institutions are turning to the Internet to provide distance learning for their students. Due to the recent advances in computer technology, it has presented new methods/formats of teaching and learning. As class sizes continue to increase, E-learning has become a powerful learning tool and will continue to play a key role in the future for teaching and learning.

These new technologies have enabled the development of new teaching methods such as virtual field trips. These are effective teaching tools for large classes. The virtual field trips also provide great opportunities for physically disabled students who would not usually be able to participate in field trips. The course at Arizona State University utilises six virtual field trips about south-western landscapes in the area.

The virtual field trip can be accessed through the Internet and it is presented as a series of web pages with a structure consisting of the index, field trip stops and concepts. The virtual field trip includes a number of photographs—both aerial and panoramic.

Approximate cost of technology:
It is a cost-effective alternative to field trips.

How much time was taken to develop this and implement the method?
The Introductory Physical Geography classes generally consist of three hours of lecture and 3 h of learner-centred laboratory work each week over the course of sixteen weeks.

What evidence is there of the effectiveness of this activity?
Numerous studies have concluded that virtual learning greatly improves field experiences when used as a preparatory tool, but they also indicate that they should not replace field trips. virtual field trips and instruction prior to the fieldwork allows students to focus on the key concepts and use their time more efficiently.

It has proved to be a successful teaching method as it is accessible by physically disabled students who would not normally be able to participate in field trips. Studies have shown that virtual field trips and real field trips are equally effective methods for teaching students basic knowledge about desert geomorphology in introductory physical geography classes. It is also a cost-effective alternative to real field trips.

Pitfalls/Problems/Limitations:
There are a number of limitations of virtual field trips. One problem is that they do not provide the same sensory experiences as a real field trip. Many scholars highlight the importance of the five senses when it comes to virtual field trips. The visual and audio capabilities of multimedia can provide almost the same sensory experiences as actual field trips by using pictures, audio clips and videos.

However, it cannot emulate smell or touch. Touch is often an integral part to physical geography, so students can experience the mass of rocks or feel the texture of soils for example. Smell is the other sense that cannot be emulated through multimedia, so students will be incapable of detecting smells. In virtual learning, students must read about these other concepts and imagine their sense of touch and smell.

Fieldwork generally promotes teamwork and co-operation through communication in the field. This is missing from virtual fieldwork, although email and discussion threads can provide interactive opportunities, it is without the expression of emotion and facial expression, without this miscommunication is common.

Although both virtual and real field trips are both effective teaching methods, virtual field trips become less effective when the difficulty of the work is increased, as greater association to the physical location is needed. Students who participated in the real field trip showed greater understanding than those who participated in the virtual field trip. So in conclusion, VFT are effective in introductory lessons but less so beyond that in their current state.

Technological expertise required:
Staff skills required: Basic computer skills needed and a good knowledge of the virtually created field trip is needed to instruct the students. Student skills required: Basic computer skills needed.

See also: Stumpf et al. (2008)

7.4 Summary

The initial hopes (for those into cost-cutting) that virtual field trips might allow students to benefit from a field experience without expense have been largely abandoned because of the time needed for construction. The re-usable object approach promises to be a useful one to help construct guide-oriented trips. The accessibility of *Google Earth* and the increased student involvement in this technology is a useful way of bringing together spatial, geospatial (GIS) approaches to task oriented or problem-based tasks. *GigaPan* technology (see Chap. 8) is much simpler and less expensive than using Lidar mapping, at least for educational purposes and once purchased, the equipment can be used as part of group activities. On an even simpler level, apps such as *Photosynth* can be used with iPad or smartphone cameras to produce simple panoramas that can be built into larger projects, perhaps with *Layar* approaches. All these hands-on technologies enhance student engagement and project ownership whether at an individual or group level.

References

Chickering AW, Gamson ZF (1987) Seven principles for good practice in undergraduate education. Washington Center News, Fall 1987, cited from: http://www.lonestar.edu/multimedia/SevenPrinciples.pdf

Curiosity Rover (2014) Mars science laboratory curiosity rover. Retrieved from http://mars.jpl.nasa.gov/msl/multimedia/interactives/photosynth/

Elvivo (2014) Virtual eruption. Retrieved from http://evivo.pi.ingv.it/

Fieldwork (2014a) Guidance and information about fieldwork for students by students. Retrieved from http://sarahnolan15.wix.com/fieldworkforstudents

Fieldwork (2014b) Selection of student videos centred on fieldwork methods. Retrieved from http://sarahnolan15.wix.com/fieldworkforstudents#!video-menu/cupt

Jorum (2014) Jorum website. Retrieved from http://www.jorum.ac.uk/

Stott T, Nuttall A, McCloskey J (2009) Design, development and student evaluation of a virtual alps field guide. Planet 22:64–71. Retrieved from http://journals.heacademy.ac.uk/doi/full/10.11120/plan.2009.00220064

NearestTube (2014) Retrieved from (http://www.nearest-tube.com/)

Polsani PR (2003) Use and abuse of reusable learning objects. Texas Digital Library 3(4). Retrieved from https://journals.tdl.org/jodi/index.php/jodi/article/view/89/88

Stimpson I (2009) Autobrecciated rhyolite, gigapan image. Retrieved from http://www.gigapan.com/gigapans/23097

Stimpson I, Gertisser R, Montenari M, O'Driscoll B (2010) Multi-scale geological outcrop visualisation: using gigapan and photosynth in fieldwork-related geology teaching. The smithsonian/NASA astrophysics data system. Retrieved from http://adsabs.harvard.edu/abs/2010EGUGA..12.4702S

Stott T, Litherland K, Carmichael P, Nuttall AM (2014) Using interactive virtual field guides and linked data in geoscience teaching and learning. In: Tong VCH (ed) Geoscience research and education. Springer, Netherlands. ISBN: 978-94-007-6945-8 (Print) 978-94-007-6946-5 (Online)

Stumpf R, Douglass J, Dorn RI (2008) Learning desert geomorphology virtually versus in the field. J Geogr High Educ 32(3):387–399

Sung D (2011) Augmented reality in action travel and tourism. Retrieved from http://www.pocket-lint.com/news/108891-augmented-reality-travel-tourism-apps

Thorndycraft VR, Thompson D, Tomlinson E (2009) Google Earth, virtual fieldwork and quantitative methods in physical geography. Planet 22:48–51. Retrieved from http://journals.heacademy.ac.uk/doi/full/10.11120/plan.2009.00220048

Virtual S (2014) Virtual Skiddaw: 3D geology field trip. Retrieved from https://learn5.open.ac.uk/course/format/sciencelab/section.php?name=skiddaw_1

Whalley WB (2010) Scrabo hill, west quarry face, gigapan image. Retrieved from http://gigapan.com/gigapans/57270

Whitmeyer SJ, Bailey JE, De Paor DG, Ornduff T (2012) Google earth and virtual visualizations in geoscience education and research. Boulder, CO, Geological Society of America, Special Paper 492

Chapter 8
Portable Networks and Specialised Fieldwork Applications

Abstract In this chapter we bring together some fieldwork-related 'extras' that are technological but not directly related to iPads or tablet technologies. Through case study examples the potential of local wireless networks are considered as well as a range of specialist apps to support student fieldwork recording and measurement.

Keywords Local wireless network · Field network system · Remote fieldwork · Fieldwork apps · Field photography

8.1 Introduction

For the most part it not envisaged that students will be let loose with the equipment mentioned in this chapter except under guidance or after training. But this depends upon your viewpoint. One is that, if students are going to work in the commercial and academic world, then they should be introduced to technology.

We also include a variety of apps that may be useful for all sorts of projects, disciplines and fieldwork. In the section on specialised fieldwork applications we consider the hardware and related software that can be used on fieldwork or to enhance the fieldwork experience for students.

8.2 Portable Networks: Remote Access Through Local Wi-Fi Services

The next two case studies by Trevor Collins and colleagues show how putting a variety of technologies together can benefit students with accessibility difficulties.

Case Study 28: The Field Network System: Using a portable Wi-Fi network at fieldwork sites without Internet access (Trevor Collins and Jim Wright)

Institution: The Open University and the Field Studies Council, UK

Keywords: Outdoor learning; fieldwork; group discussion; peer dialogue; co-operative learning; inquiry learning.

Aim: To develop a Field Network System (FNS) which can facilitate the collection, collation and interpretation of group datasets at fieldwork sites where Internet access is unavailable or unreliable.

Outdoor learning provides exposure to learning environments that promote authentic practical inquiry and complement classroom learning. Fieldwork, in particular, seeks to exploit the affordances of place and group work, in order to: contextualise the learning experience, facilitate individual growth and support the development of social skills. However, data collection and interpretation are often separated in both space and time: data collection is typically carried out at a fieldwork site, whereas data collation and interpretation is undertaken back in a classroom or lab. For learning activities, in which, group datasets can be collated and plotted automatically, the FNS can help guide the data collection process, and support the students' discussion and interpretation of their results while they are at a fieldwork site.

The system was created through a collaborative project involving the Field Studies Council (FSC) and the Open University (OU). The FSC is an environmental education charity, based in the United Kingdom that was established in 1943. The OU began working with the FSC to create fieldwork courses and project guides in 1973 and the FNS project was initiated in 2013, as part of the OU's OpenScience Lab (see OU 2013).

The FSC's inquiry-based learning activities typically start in a classroom with a briefing to introduce a topic, develop the research question or hypotheses, discuss and agree the method of data collection, and explain the use of any associated equipment. The class of students, schoolteachers and FSC tutors then travel to a fieldwork site where they collect their data (typically on paper). After a few hours at a site, they return to the field centre and enter their data into spreadsheets, which are checked and corrected where necessary, and collated into a class data set. Graphs or charts are plotted, so that students can interpret their results and discuss their findings. The students work in small groups throughout their time at a field centre.

The level of student autonomy and support provided by tutors varies according to the students' age range and curriculum, but the form of authentic practical inquiry is common. The learning activities are aligned with the national curriculum for schools and customised to accommodate the requirements of the corresponding exam board. The accessible fieldwork sites in the geographical area help determine the portfolio of courses available at each of the FSC's field centres. The FNS was initially developed with the education team at the residential centre in Preston Montford (near Shrewsbury, Shropshire). The website was developed to support three popular lesson plans for schools visiting that centre, namely: a pond trophic structure study, a river discharge study, and a pond wet system study. The system is now being refined and extended to meet the needs of the education teams at other FSC centres.

Technology required:
The FNS includes a battery-powered Wi-Fi router and laptop, which are used to create a local network and website that students can access using any web-standards compliant browser on a mobile device (such as a Smartphone or tablet, see Fig. 8.1). As noted above, the website is tailored to support a number of inquiry-based lessons that the FSC run for school groups. The website is managed through an open-source content management platform, so that the site can be extended to support additional lesson plans using a series of web forms. The 'theme' used by the content management platform to display the website uses a 'responsive web design' approach, which modifies the style and layout of the page dynamically in response to the dimensions of the browser window. This ensures that the site can be accessed effectively on any device using a web-standards compliant browser.

As evidenced in this publication, mobile technologies offer an effective set of tools for structuring and guiding fieldwork learning. However, without reliable network or Internet connectivity, mobile applications are constrained

Fig. 8.1 The field network system (FNS), comprising a small rucksack, battery powered Wi-Fi router and laptop (*left*), which are used to create a local network and website to support inquiry-based fieldwork learning activities using a web-standards compliant browser on mobile devices (*right*). Photographs courtesy of Trevor Collins

in the level of support they can provide for social group interaction, which is a core aspect of fieldwork learning. Therefore, less populated, rural and remote fieldwork locations are unlikely to have 4G/3G Internet access. As a result, portable network solutions are needed in locations where Internet connectivity is unavailable or unreliable, in order to provide a basis for using mobile technologies as effective collaborative tools.

The distribution of a group of students in the fieldwork environment should be considered when selecting the network Wi-Fi router. More specifically, the coverage of the router's antenna needs to be taken into account and matched to the geographical distribution of students. Antennas are typically either directional or omnidirectional (i.e. they either send and receive signals in a directed beam or in all directions). Directional antennas are good in beach locations, where students are spread along a flat area. Omnidirectional antennas are better suited for pond locations, where students work within an area. In locations where all the students are not within sight of a single point (i.e. the location of the Wi-Fi router) multiple routers can be used to extend the range of the network. In practice, FSC students at fieldwork locations are usually within sight of a tutor, therefore, in most cases a single router with an omnidirectional antenna is used to provide a sufficient network.

With regard to mobile applications, the FNS is an example of a web application, in that it uses a web browser to access an entirely web-based service. An alternate approach would be to produce a native or hybrid application. A native application runs entirely on the mobile device, where as a hybrid application uses an application running on the mobile device to access web-based content. The main benefit of using a web application is that no software needs to be downloaded and installed on the mobile devices, because a web-standards compliant browser can be used to access the website. Note there is an important distinction to make between the web and the Internet: the web refers to a system of interlinked hypertext documents and the internet refers to the global system of interconnected computer networks. The FNS is a web application that can be used in locations without Internet access.

Approximate cost of technology:
Table 8.1 lists some examples of the technology used and indicative costs. An FNS kit comprising a laptop (£350), Wi-Fi router (£80) and battery (£70) currently costs around £500. Laptops with solid-state drives (SSDs) are generally more robust than laptops with (standard) hard drives. Where possible, preconfigured equipment spares should be carried as swappable replacements. Any device with a web-standards compliant browser can be used with the FNS.

How much time was taken to develop this and implement the method?
The FNS was initially developed during a six-month project as part of the OpenScience Lab, and builds on previous work at The Open University on

Table 8.1 Examples of the technology used for the field network system and indicative costs (from 2013)

Item name	Example models	Price (inc. Tax)
Laptop	Samsung ATIV Book 9 Lite Laptop, Quad-core Processor, 4 GB RAM, 128 GB SSD, 13.3″, White	£500
	Acer Aspire E1-530 Black 15.6″ PDC 2117U 4 GB 500 GB DVDRW Win8 64-bit	£350
Wi-Fi router	Ubiquiti NanoStation M2 (directional antenna)	£90
	Ubiquiti PicoStation M2 HP (omnidirectional antenna)	£80
	Ubiquiti NanoStation Loco M2 (directional antenna)	£50
Battery	External laptop battery (20,000 mAh)	£70
	External battery pack (15,000 mAh)	£40

fieldwork and mobile learning. During that time the system was created, along with website support for three FSC lesson plans.

The Wi-Fi network routers take approximately an hour to configure. Depending on the services being used, the laptop web server software installation and configuration takes around half a day. The router settings can be saved as configuration files and used to set up additional routers of the same type. Similarly, for the laptop, once the web server software is installed and configured, a copy of the laptop's configuration can be saved as an image file and re-used to configure additional laptops of the same specification. The main benefit of reusing configuration files and installation image files is that it ensures identical kit configurations and minimises opportunities for human error.

The majority of the time taken to develop the system was spent in tailoring the website to support the FSC's lesson plans and school group bookings. For a given lesson plan the site supports data entry (and editing) using web forms for task-specific content types, and data collation and presentation using table and chart views. As with any website, additional content types and views are needed to support further lesson plans. Although, parts of the website design may be re-used, additional lesson plans are developed iteratively in response to tutor and student feedback, and generally take around one month to develop.

What evidence is there of the effectiveness of this activity?
During the development of the system, trials with tutors, teachers and students were observed; several tutors were interviewed; and comments and feedback were collected and used to revise the system. In particular, the feedback from the tutors involved in the development iterations justified the commitment from the FSC to use system at their Preston Montford centre and deploy it in more of their centres. As the system becomes part of the FSC's usual teaching practice, the on going work with the university will focus on evaluating the impact on the students' fieldwork experience, particularly with regard to the focus and structure of their discussions at a fieldwork location.

Pitfalls/Problems/Limitations:
Limited Wi-Fi range and antenna coverage:

- The Wi-Fi routers have a limited range, therefore, the coverage of the routers was tested at fieldwork locations and the tutor's selection of recommended equipment is based on common scenarios, specifically: ponds (i.e. a Ubiquiti PicoStation), beaches (i.e. a Ubiquiti NanoStation) and rivers (i.e. multiple routers, if necessary).

Technology-failure:

- Assume the technology will break and prepare accordingly. Carrying swappable spares (i.e. a spare battery, router and laptop) minimises any resulting delays at the fieldwork site.
- Assume that the website won't do everything the students need, so make sure that they can also access their data in a standard format (i.e. as comma separated value 'csv' files) that can be opened with a spreadsheet application for further analyses.
- Time at a fieldwork site is limited, therefore, have a set of simple diagnostic steps that a tutor or student can carry out if something is not working. Namely, checking the router and mobile device connection, checking the website is running, and checking that the web address is entered correctly in the browser on the mobile device.
- If restarting a device does not fix the problem, then it can be replaced with a spare, failing that a paper-based alternative is used (i.e. data sheets and pens).
- Carrying a paper-based alternative to the FNS provides a reliable safety net that ensures the students' fieldwork experience is not ruined by unforeseen technology problems.

Technological expertise required:
Staff skills required:

- Tutoring—Web browsing and data entry using web forms (no training required).
- IT—Network router configuration (e.g. AirOS, DD-WRT), and web server installation and configuration (e.g. XAMPP).
- Web development—Website development using a content management platform (e.g. Drupal 7).

Student skills required: Web browsing and data entry using web forms (no training required).

Key tips:

- The purpose of the system is to facilitate good fieldwork conversations, so make that the success criteria. A holistic approach is needed to ensure that

the applied technology and pedagogy are complementary. The tutoring, IT and web development skills needed to develop such a system generally requires teamwork. Making sure that everyone is focused on developing good fieldwork conversations helps unify the different approaches across the team.
- Work out what the students need to see in order to prompt them to discuss their findings and make sure the website delivers it.
- Analyse the activities students already do during fieldwork by breaking each activity down into the tasks, actions and operations they perform. Then try to use the technology to do the boring bits like collating data and plotting charts, so that the students can focus on the educationally useful activities like interpreting the results and discussing their findings.
- Develop the system iteratively, the website is unlikely to be correct first time, so have a go and try to get feedback. Take a few iterations to get feedback from a range of people before trying it with students. Never underestimate the benefits of trialling the use of the system before deploying it with students on a course.
- Always be prepared for something going wrong, it will eventually, but if there is a contingency plan then disasters are usually avoided. At the very least, always carry a paper-based alternative.

Also see: FNS (2014)

Case Study 29: The Enabling Remote Activity (ERA) project: Using a portable Wi-Fi network to enable remote participation in fieldwork for mobility impaired students (Trevor Collins, Sarah Davies and Jessica Bartlett)

Institution: The Open University, Milton Keynes, UK

Keywords: Outdoor learning; remote fieldwork; accessibility; inclusive teaching.

Aim: To enable mobility impaired students participate in fieldwork.

The Open University (OU) is a distance learning university based in the United Kingdom. The purpose of an OU fieldwork course is to help students contextualise and apply their knowledge of a subject, and develop practical discipline-based skills. A typical fieldwork course will involve students working in small groups, supported by a fieldwork tutor. The students practise relevant fieldwork skills over successive days at a range of fieldwork locations within a few hours drive of their residential centre, in order to build-up their experience and confidence. During the course, formative assessment of field notebooks and group discussion, for example, provides opportunities for the students to get frequent and immediate feedback on their work. Following a typical fieldwork course, students write a report that is marked by their lecturer as the summative course assessment.

Where possible, accessible locations are used for fieldwork courses. However, depending on a student's individual requirements, some fieldwork locations cannot be reached by mobility-impaired students. In order to access such locations remotely, a temporary Wi-Fi network is used so that a remote student, or group of students, supported by a tutor can communicate directly with a tutor and/or the rest of their group at a nearby fieldwork location in order to complete the course activities. This approach is intended to help mobility-impaired students participate in fieldwork courses without adapting the course content or student activities. OU fieldwork activities typically engage the students in problem-based learning. For example, students could be introduced to an area and encouraged to develop an explanation of how the area was formed or how an environment is changing. The problem-based approach involves the tutor acting as a guide and facilitator to support the students. Helping them explore alternate perspectives through developing and testing their own hypotheses.

The Enabling Remote Activity (ERA) toolkit allows the remote and field-based groups to communicate using VoIP (i.e. Voice over Internet Protocol) phone calls, live video streams, and photo sharing (see Fig. 8.2). Of these, the most important service is voice communication, as a significant portion of the learning process in this case is supported through discussion. Live video streaming enables both parties to see what is currently happening and can be used effectively by the remote group to help direct the exploration of a fieldwork location. High-resolution photos provide an effective way of capturing and sharing relevant fieldwork features, and can also be used as a basis for making fieldwork sketches, through which the students develop

Fig. 8.2 The ERA toolkit in use on the OU's environmental change course at Howick Haven, Northumberland in 2009. The students (*left*) are discussing the site's geology with the tutor (*right*) while watching a video feed from the tutor's helmet camera. Photos taken by the tutor are displayed on the students' second screen (A video clip is available at YouTube 2014). Photographs courtesy of Chris Valentine and Trevor Collins (both were first published in Collins et al. 2010)

important recognition and abstraction skills. For discipline-specific physical fieldwork skills, such as using a hand-lense to identify minerals or plants, providing the students with previously collected samples to work with individually, is generally more effective than a technology mediated alternative.

The ERA toolkit was used on the following OU courses: 'Ancient Mountains' (OU course code SXR399) at Kindrogan Centre in 2006 and 2008; 'Environmental Change: The record in the rocks' (SXR369) at Durham in 2009; and will be used on the 'Practical Environmental Science' course (SXE288) at Preston Montford and Malham Tarn in 2014.

Technology required:
Ideally, all fieldwork course locations would be accessible and there would be no need for the ERA toolkit, but the choice of educationally rich fieldwork locations and ease-of-access is a practical trade-off. Therefore, creating a temporary communications link is necessary in order to enable some students to remotely access inaccessible locations. Creating a local area network enables communication between devices on the network, and more specifically, between the users of those devices. Wi-Fi is an established collection of communication standards designed for creating local area networks, and is commonly used by personal and mobile computing devices, as well as by some video and photo cameras.

In locations where both field and remote groups are within sight of each other, a single Wi-Fi router or pair of routers are used to create a local Wi-Fi network. In locations where the two groups cannot see each another, additional routers are used to extend the range of the network and create indirect links. These are typically needed to create a local network around a coastline, over a cliff or over a mountain peak. Routers with directed antennas send and receive signals in a focused beam and are most effective for creating point-to-point links. Routers with Omni directional antennas send and receive signals in all directions (i.e. 360°) in a plane perpendicular to their antenna's axis, and are most effective for connecting devices within a local area. A set of six routers is sufficient for all of the fieldwork locations used in our courses. At the majority of our fieldwork locations, only two or three routers are used. Photography umbrella swivel connectors are used to mount the routers on photography lighting stands, which are extended vertically to avoid obstructions and thereby maximise the Wi-Fi coverage of each router.

The students' fieldwork activities determine the quality requirements and use made of video streams and photographs. Generally, video and photo cameras producing higher resolution video streams and photo files are more expensive, and there is a corresponding trade-off between quality and price. Similarly, higher resolution video and photo cameras create more data, which can affect the amount of bandwidth they use on the network. Therefore, the data format is important. Most digital photo cameras have adopted the JPEG format as a standard. Although alternate video compression formats are available, the H.264 standard offers very good data compression and performance for streaming video over a Wi-Fi network.

A web server application stack (i.e. XAMPP) is installed on the laptop to support file sharing and to provide a familiar website interface for students and tutors to view the video streams and shared photos. A script running on the web server (i.e. the laptop) automatically creates a photo gallery from the photo files received and creates pages for viewing the available video streams. As a result, the students and tutors use a web-standards compliant browser to access and view the photo images and video streams on a tablet or laptop computer connected to the local network.

Although the ERA toolkit requires a significant amount of prior configuration, the kit is designed so that fieldwork students and tutors need very little training to operate it. Providing the routers are placed in appropriate locations, the tutor only needs to connect the batteries to the routers and switch on the laptop. All the software services are started and stopped automatically, when the devices boot-up and shutdown. Shortcuts to the website that display the photo gallery and video streams are added to each mobile device, and the same address is set as the homepage on the devices' web browser. The students generally find the website intuitive to use. Although some students and tutors are unfamiliar with VoIP client applications, most are familiar with Skype and have no problems using the phone service.

How much does it cost to implement the method?
Table 8.2 lists some examples of the technology used and the associated costs. An ERA kit costs around £2200, comprising: a laptop (£500), two tablets (£400), an IP video camera (£180), six Wi-Fi routers (£360), six batteries and connectors (£270), six stands and connectors (£330), and weather-proof bags (£160). Laptops with solid-state drives (SSDs) are generally more robust than laptops with (standard) hard disc drives. Where possible, preconfigured equipment spares are carried as swappable replacements. Any device with a web-standards compliant browser (e.g. tablets and Smartphones) can be used on the local network although additional plug-ins or extensions are sometimes required to view the video streams.

How much time was taken to develop this and implement the method?
The ERA toolkit has been developed iteratively since 2006 through use in successive courses. The majority of the development work took place in 2009 and 2010, with funding support from JANET and JISC. The initial toolkit used two-way radios for voice communication, a wired IP camera for video, and photos were uploaded to a local web server periodically via a USB cable connection. The Wi-Fi photo camera was introduced in 2007, along with the Wi-Fi video camera, and the VoIP service was added in 2009.

To recreate the current toolkit would take about one day to configure the equipment. Configuring the set of six routers is done by entering the settings for

Table 8.2 Examples of the technology used for the enabling remote activity project and indicative costs (from 2013)

Item	Example model	Price (inc. Tax)	Quantity	Cost
Laptop (server)	Samsung ATIV Book 9 Lite laptop	£500	1	£500
Tablet (clients)	Google Nexus 7	£200	2	£400
IP video camera	Axis M1004-W network camera	£180	1	£180
Wi-Fi router	Ubiquiti PicoStation M2 HP	£80	2	£160
	Ubiquiti NanoStation Loco M2	£50	4	£200
External battery	External battery pack (15,000 mAh)	£40	6	£240
Battery connectors	Power over ethernet (PoE) Injectors	£5	6	£30
Router stand	Compact photography light stand	£35	6	£210
Stand connector	Photography umbrella connector	£20	6	£120
Rucksack	Berghaus 24/7 25 l day sack	£25	4	£100
Drybag	Exped fold drybag (small)	£10	6	£60
			Total	£2200

a wireless distributed system into one router (using the router's web interface) and then saving a configuration file, which is re-used to configure the other five routers. Re-using the configuration file ensures that the routers are set-up identically and minimises the potential for operator errors. The web and VoIP services take about half a day to install on a laptop. As with the routers, once the laptop is set-up the configuration is saved as an image file that can be re-used to set up subsequent laptops with a similar specification. The scripts used to create the web interface can be downloaded from the ERA website.

Prior to deploying the system, a half-day training session for the tutor provides an opportunity to explain how to operate the equipment, and helps the tutor adapt their tutoring approach so they can work effectively with a remote group. Specifically, this ensures that the tutor knows which routers to use at each fieldwork location, and how to coordinate the use of video and photo images, in order to facilitate the discussion and ensure the learning objectives for the course are covered.

What evidence is there of the effectiveness of this activity?
In 2011 an evaluation study was completed involving two lecturers, a demonstrator and 41 students from Plymouth University (Stokes et al. 2012). To compare the direct and remote fieldwork experience, a half-day fieldwork activity was run four times with around a quarter of the students at a disused arsenic mine. The purpose of the fieldwork was for the students to work in small groups to complete an environmental impact assessment of the site being mined again. For two of the sessions, the students toured the mine supported by a lecturer. In the other two sessions, the students completed the same activity using a remote link from the car park at the mine to the tutor in the mine, who gave the same tour. No significant differences were identified between the direct and remote groups in relation to task performance, or in relation to the students' self-rated competencies in a variety of skills before or

after the fieldwork. Although, in post-fieldwork focus group discussions, the students expressed a preference for the direct fieldwork experience, they also identified a range of situations in which remote access could enhance fieldwork provision.

Pitfalls/Problems/Limitations:
Bright sunlight: The most challenging problem in using the ERA toolkit has been viewing the screens in bright sunlight. Clear plastic sample bags and dry bags are used to protect the equipment from wet weather, but other than avoiding direct sunlight on the screen it is very difficult to solve the sunlight problem. Umbrellas and sunshades have been used to shade the screens, and an external monitor can be used with laptops and some tablets. External monitors are generally better in bright sunlight than most laptop and tablet screens, but they do not solve the problem totally.

Voice communication: VoIP telephony requires the timely delivery of data across the network; any delays or loss of data will result in the speech breaking up, which makes the system unusable. As routers are added to extend a network, the potential for delays and loss of data increases. Therefore, the network configuration needs to optimise the timely delivery of voice data. Our solution is to use the minimum number of routers. Where multiple routers are necessary, pairs of routers are used to create point-to-point links. Multiple point-to-point links are operated on separate Wi-Fi channels (i.e. non-interfering radio frequencies), and are connected using an Ethernet cable between one of the routers in each link to form a chain. This solution maximises the quality of the Wi-Fi links and minimises the delays introduced when the routers switch between sending and receiving data.

Technology mediated conversations: Even though the system supports synchronous communication, the technology mediates the discussion, which inevitably influences the flow of conversation. The subtle responses of the people involved are less apparent than in face-to-face conversations. As a result, tutors need to pay close attention to the students' responses and reactions, and adapt their conversation to ensure that they are able to gauge the students' understanding and work with the group to help them all participate effectively in group activities.

Technological expertise required:
Staff skills required:

- Tutoring—Web browsing and phone calls (no training required). Setting up the stands and switching on the kit (half a day training).
- IT—Network router configuration (e.g. AirOS, DD-WRT); web server (e.g. XAMPP) and VoIP server (e.g. Asterisk, Mumble) software installation and configuration on the laptop; VoIP client application installation and configuration on the mobile devices (e.g. Sipdroid, Mumble client); and website server scripts (available from the ERA website).

Student skills required: Web browsing and phone calls (no training required).

Key tips:

- The purpose of the system is to enable students that cannot access a fieldwork location to explore it remotely from a location nearby (ideally within sight). Using the kit should be avoided unless it is necessary, where possible, help any mobility-impaired students get to the location with the rest of their group.
- When using the kit, ensure that the students have an active role in guiding the fieldwork activity. The system provides remote access to inaccessible locations, but like any other form of fieldwork learning, the students' active engagement in the process is essential in order to facilitate their learning.
- Always be prepared for something going wrong and develop appropriate contingency plans so that disasters are avoided. For example, carry spare preconfigured equipment to replace faulty equipment, and prepare an alternate experience in case a problem with the technology cannot be resolved. In geology, previously collected rock samples give students experience using hand lenses and are therefore used routinely, however, if necessary, the tutor can also use the rock samples to create an alternate experience.

Also see: ERA (2014a, b), Collins et al. (2010), Gaved et al. (2006), Lea and Collins (2009) and Stokes et al. (2012)

8.3 Photography

In this section we consider how photography can be used to enhance the fieldwork experience from the use of microscopes, telescopes, DSLR cameras with a long-focus zoom lens to create a detailed *GigaPan* images through to the versatile use of the Hero camera.

8.3.1 Microscopes

It is possible to use relatively high-powered optics for microscopic work. The *ProScope* system has been available for several years as a USB fitting to desktops and laptops. This is still available at Proscope (2014). More recently, the system comes with a body that uses a local Wi-Fi to interact with up to 250 iPads. This

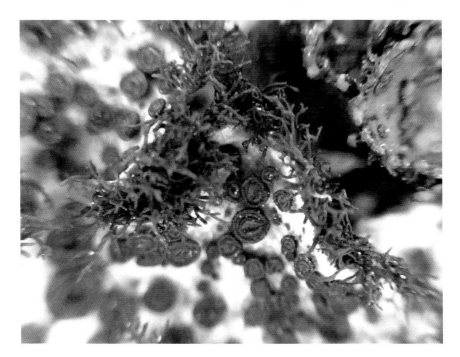

Fig. 8.3 Photograph of a lichen on a rock taken in the field using a *ProScope* attached to an iPad Mini

enables images to be shared or investigations to be done in groups. Although it can be hand-held this is clearly not possible for magnifications above about ×50. In this case a platform and stand are needed for field use. The device can take video and scaling and annotations can be used on captured images. Other devices are available but the *ProScope* seems to offer good flexibility with objectives being interchangeable (Fig. 8.3).

8.3.2 Telescopes

For higher magnification as a telescope, there are now fittings available to link a mobile/smartphone to a spotting scope, binoculars, night-vision optics, telescope or bench microscope. One of these with a range of applications is *PhoneSkope* (see PhoneSkope 2014). These clearly have research and teaching as well as hobby applications. For high quality use, DSLRs can be fixed to various instruments in a similar way and images downloaded via a card reader.

8.3.3 GigaPan

The various *GigaPan* instruments allow a conventional camera to take multiple overlapping images to be taken. It consists of a frame to hold the camera on a rigid base, usually a camera/video tripod (Fig. 8.4). Sizes of the frame vary according to size of the camera, which can be large DSLR and lens, or small 'pocket' camera. The device is programmed for a variety of menu-driven options. The focal length is set fixed (for a zoom lens) throughout the image taking session. Location upper left and lower right corners of the field seen through the camera set the coverage of the required field. A full 360 panorama can also be taken. Once the stepping motor driving the sequence is started, the overlapping vertical horizontal images taken are stored in the camera memory. These images are downloaded at base, collated, processed and uploaded to the *GigaPan* site. The upload is also accompanied by metadata, such as a site description, direction altitude. The images are essentially digital resources that others can also repurpose.

Examples of *GigaPan* images on the web have been mentioned in Chap. 7. As images are being added all the time, a search for 'gigapan geology' will provide a wide variety of sites and applications.

Figure 8.5 shows a *Google Earth* image taken as a screenshot, with added text to show a landslide feature. The *GigaPan* associated with Fig. 8.5 is available at Whalley (2014) UTM co-ordinates will provide the location and of course a measure of oblique viewing is possible using the controls on *Google Earth* but this does not usually provide anything like the detail that can be obtained by a terrestrial *GigaPan* view. Used together with a variety of additional questions, quizzes,

Fig. 8.4 DSLR camera with long-focus zoom lens on a *GigaPan* frame

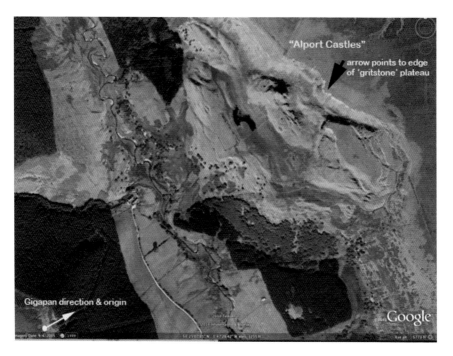

Fig. 8.5 Annotated *Google Earth* vertical image with the origin of a *GigaPan* view. This could be built into a website to provide more information or perhaps a set of associated questions. (Map data: Google, Tele Atlas)

instructions, perhaps on a website, *GigaPan* views can provide an effective way of adding detail to aid students' exploration of an area.

Technical Note:

It is possible to add text, arrows to a *GigaPan* image so that it becomes visible in a close-up. However, this is not easy as it is necessary to select the image to be displayed of the perhaps hundred that make up the composite, add the text and reinsert it and upload the whole image batch again to the *GigaPan* site.

8.3.4 Hero Camera

The Hero series of cameras are small, compact, cameras storing images on an SDHC card (Fig. 8.6). The camera can be installed in waterproof housing, which can be body, helmet, vehicle or tripod mounted. Because of this housing, the cameras can take wide-angle views in single images or video under adverse weather conditions or, for example, underwater. A variety of modes allow delayed action, time sequences etc. Although there is no viewfinder, an optional video screen on the

Fig. 8.6 The *Hero* camera (*right*) with SDHC card, *centre*, waterproof case and *left*, bracket and camera tripod fitting. Photo using the on-board camera of an iPad. The card's images (.mov) can be read from the card through a card reader onto the iPad and processed as required from an app such as *Splice*

back shows the view and a Wi-Fi link is also available. This is professional quality and a favourite with active sports video-photographers so can be used in a variety of adverse conditions, including underwater video, which may be useful for biological applications.

Hero cameras and equipment are been added to and upgraded, to include remote operation and mountings. These add further fieldwork possibilities.

8.4 Surveying Equipment

Many types of fieldwork involve some sort of surveying. This might be added information for a project, such as a transverse profile or a small topographic feature where more resolution is needed. A surprising amount can be undertaken using and iPhone or iPad and we mention a few of the appropriate apps below.

Distance measurement is not (yet) included in any tablet but all have a spirit level or clinometer of some kind. Practice shows that these are easier to read than clinometers (which may cost £100), have a 'hold' button, show the angle (in degrees, decimal, percent or 'x in y') with selectable accuracy and can even speak the value.

Distances in field surveying in the earth sciences have traditionally been done by flexible tape measure. However, these are difficult to use in a wind and do not take kindly to sand and water. Simple laser measurement devices, under various trade names, can be found in hardware stores. The range of the laser depends upon the price and a little skill is required to see the red laser spot. However, they are easy to use and can be put in a pocket.

Two other useful apps that can be used for simple surveying also use the clinometer ability. *SeeLevel* can measure distances by using some local standard such as a metre rule or the height of a person. *Theodolite* is a little more complex but is a versatile app for field surveying (Fig. 8.7).

We should also consider the use of professional surveying devices for use with student groups. The main instruments here will be differential global positioning system (DGPS), digital total station theodolites (see for example in the case study in Chap. 5 by Ian Fuller) or levels. Usually, the instruments will have and require sophisticated data reduction applications that will need to be processed at base. Nevertheless, the results could be processed and placed into a report, such as the surveyed river sections inserted into a *Google Earth* image in Fig. 8.8.

In the last few years microprocessors and good battery life have brought some surveying instruments to what might be called a 'semi-pro' version such that students can use them easily for project work. As with other aspects of fieldwork, they give a rapid way of taking measurements, both distance (with a laser) and angle/clinometer. Thus two and three-dimensional measurements can provide area and volume calculations 'on board'. With the robustness and water and dust resistance (IP54) they make ideal field measurement devices for students. They

Fig. 8.7 Screen shot of an iPad photograph with the overlaid information from the surveying app *Theodolite*

Fig. 8.8 A *Google Earth* image with DGPS-surveyed profiles of a river (under the tree canopy). This image was shared between the groups doing the surveying. (Map data: Google)

allow authentic problems to be solved in a modern way and, in our view, should replace optical clinometer and tape measures. Not least because they are cost-effective and are professional devices for surveyors, foresters, site engineers etc.

Some of these devices can interface directly to an iPad. Even if this is not done, the *DISTO Sketch* app allows calculations to be done by direct entry and also allows importing of an image that can then be annotated with measurements taken etc. Measurements can be added to images and sketches and these shared as necessary. We have not yet seen this implemented in student fieldwork but it is another example of the diverse ways in which iPads can be used in professional situations.

8.5 Specialised Fieldwork Apps

In this section we list a few pieces of fieldwork kit that may prove useful when students need to make measurements in the field. They add functionality to a tablet or Smartphone and are, for the present, not bundled with tablets. They are included in the usage schema of Chap. 1, Table 1.2.

8.5.1 Distance and Angle Measurement

Traditionally, distance measurements for fieldwork are taken using flexible tapes ranging from 2 m up to 50 m. While the former are useful to have in a pocket, are very cheap and easily replaceable, the latter are not. Experience shows that they get clogged with sand and mud, abrade over sharp rocks and are difficult to handle in a wind. A better idea is to use a laser distance metre. Although some of these are sophisticated (e.g. *Leica's Disto* range) and can include clinometers, an off the shelf one can be bought for less than £20. However, there is usually a range—price trade off and also in accuracy. The latter is well up to that provided by a tape, especially if there is a wind and, unlike a tape in catenary, does give a straight-line distance!

Having got this rangefinder it is possible to take measurements solo, without needing a fellow student or getting in difficult or dangerous locations. It is also much more rapid. There are several apps that allow the single observer to get the height of trees, profile slopes etc. We have used, in increasing sophistication:

Clinometer: provides a basic angle measurement in degrees, percent, x in y and with a speech synthesis to call these readings out. It also provides a level, to help you put up a shelf for example! You can also select the required accuracy. This basic app does away with the need for expensive clinometers.

SeeLevel: providing a clinometer/'sextant', level, parallel (stadimeter) and slope distance measurements. With a bit of geometry all sorts of measurements can be taken.

Theodolite: this provides a range of measurements that emulate a proper surveying theodolite with the additional ability to relate to a map (Via *Google Earth*), make notes and take photographs that can be mailed. This is quite a sophisticated device and can make distance measurements without an independent rangefinder. Some practice is needed before going into the field with this app.

All these apps need calibration but this is quick and easy to perform.

8.5.2 Geological Angle Measurement

Geologists may need to take various sets of angular measurements in the field; dip, strike, plunge and these need to be associated with a compass bearing. These are usually done by a compass-clinometer (Brunton compass in the USA). These readings, and many are required at an outcrop, are most easily done with two people, one taking the measurement and the other recording the measurements. Three currently available apps provide the ability to take the measurements with a tablet/smartphone and then record and even plot them on screen.

Lambert: This takes its name from a Lambert polar co-ordinate project (although Schmidt and rose diagrams can be plotted). Notes can be taken and it is possible to e-mail sets of results. A manual setting means that students might like to experiment in the lab to get readings off a book spine before tackling plunging anticlines on a windswept crag!

GeoID: This provides similar functions to *Lambert* but with the addition of map-located data collection. CSV file collection allows compilation of results off site.

FieldMove: This does much the same at the two previous apps but with the useful facility of showing a digital compass at the same time as the clinometer, useful for lineation mapping.

They will be invaluable for individual or group projects, especially when students are learning their trade. We have tried them out but not, as yet, built them into any fieldwork.

8.5.3 Light Measurement

LuxMeter and *LightMeter* are two that can use either front or rear cameras to obtain a light measurement (in lux) with a 10 s timer. This might provide an experiment in leaf sunlight absorbance.

8.5.4 Temperature Measurement

iCelsius: tablets do not have temperature sensors. *iCelsius* has a waterproof sensor on the end of a thin wire leading to an A/D converter plugged into the tablet port. A scalable and rapid response continuous graph is produced.

8.5.5 Magnetic Field Measurement

Magnetic Field: unsurprisingly, this gives a measure of the local magnetic field (µT) by way of the internal magnetometer and a direction arrow showing inclination.

8.5.6 Sound Measurement

With its audio in/out facility it is not surprising that there are several apps for sound measurement, some of these (e.g. *Decibel*) are free. These apps might be used to compare traffic noise in a neighbourhood or presentation applause. Peak/maximum values can be shown and sampling rate and calibration are usually shown.

8.5.7 Time Measurement

There are many apps working as timers and clocks. These might have uses from home baking to event timing and laboratory procedures as well as wake-up for breakfast! Usually, such timers will work behind the scene while other operations occupy the tablet.

8.5.8 Accelerometer

Most tablets have accelerometers and a basic use of these for fieldwork would be for detecting seismic activity. You would not want to occupy a tablet on this function perpetually but it might be useful in a demonstration of wave types.

8.5.9 Calculators

Although digital calculators are almost disappearing as hardware items, the original Hewlett-Packard HP 41C calculator introduced in 1979 was the first with an alphanumeric display. In effect it was the first desktop computer. All sorts of tablet-based calculators now exist and are still useful for doing calculations. Their use in fieldwork is still important.

8.5.10 Geoscientists

Geoscientists have quite specific needs for fieldwork. Not only recording and field sketching but also the need to take multiple angular measurements associated with exposed geological structures. Traditionally, these measurements were written in field notebooks and plotted on stereographic nets to show 3-D structures in two dimensions. Apps such as *Lambert* allow the plots to be recorded in the field (along with a measure of the earth's magnetic field at that point (in μTesla). This ability is especially useful for teaching in the field and showing students this important skill. *GeoID* is a similar app with the ability to geoposition and show (when Wi-Fi connected) the *GoogleEarth* Location. A more sophisticated field notebook, *Fieldmove*, also provides the clinometer and GPS facilities and links to appropriate imported base maps.

SedMob is an app (currently only for Android) that allows sedimentary log data to be entered in the field and graphic log diagrams to be constructed (Wolniewicz 2014). Customisation and export facilities are provided and this app provides a means of getting field notebooks into an electronic format.

Glacial striae (scratches) and sediment fabrics are measurements relying on a clinometer used by geomorphologists and sedimentologists. At present however, we do not know of an app that will enable these to be measured and plotted on a polar diagram or with a statistical analysis such as eigenvector analysis.

8.5.11 Species Identification Guides and Citizen Science Apps

There are now many species identification apps available and more guides available via interactive websites. These can assist with species identification in the field and often the apps offer a field notes functionality to allow recording of your own geo-tagged sightings and photographs. Field biodiversity identification is also supported by Web 2.0 initiatives such as *iSpot* which is an online community that helps identify wildlife and share nature and the *Opal* (Open Air Laboratories) network which is a UK-wide citizen science initiative. There are many opportunities for students to contribute towards large research projects by collecting and sending in field records to citizen science projects. There is an ever-expanding list of these apps and websites at our *Pinterest* site (see E-fieldwork 2013).

8.6 Summary

Improvements in battery technology as well as computing devices mean that it is possible to have Wi-Fi links between devices in the field. Students (and perhaps tutors) with accessibility problems will certainly be able to benefit by making what is essentially an 'outside broadcast'. This gives an immediacy of experience that is not possible by just videoing an area of interest or making a virtual field trip. The additional functionality offered to tablet or Smartphone by the range of specialist apps highlighted in this section demonstrate the diverse measurement opportunities available to students in the field. These opportunities can only increase in the future with the development of additional apps.

References

Collins T, Gaved M, Lea J (2010) Remote fieldwork: using portable wireless networks and backhaul links to participate remotely in fieldwork. In: The proceedings of the 9th world conference on mobile and contextual learning (mLearn 2010), 19–22 October 2010, Valletta, Malta. Retrieved from http://oro.open.ac.uk/24711/

E-fieldwork (2013) Enhancing fieldwork learning project pinterest site. Retrieved from https://www.pinterest.com/efieldworkl/

ERA (2014a) Project website. Retrieved from http://projects.kmi.open.ac.uk/era

ERA (2014b) Project blog. Retrieved from http://projects.kmi.open.ac.uk/era/blog

FNS (2014) Project website and blog. Retrieved from http://weblab.open.ac.uk/fns

Gaved M, McCann L, Valentine C (2006) ERA (enabling remote activity): a KMi designed system to support remote participation by mobility-impaired students in geology field trips. KMI technical report: KMi-06-15. Retrieved from http://kmi.open.ac.uk/publications/pdf/kmi-06-15.pdf

Lea J, Collins T (2009) VoWLAN toolkit technical report—JISC rapid innovation project report, December 2009. Retrieved from http://projects.kmi.open.ac.uk/era/vowlan/2009/11/30/portable-vowlan-a-portable-voice-over-wireless-local-area-network-for-mobile-learning/be better?

OU (2013) The OpenScience laboratory website. Retrieved from http://learn5.open.ac.uk

PhoneSkope (2014) An optics application website. Retrieved from http://www.phoneskope.com/

Proscope (2014) Proscope accessories, information and resources website. Retrieved from http://www.bodelin.com/proscope

Stokes A, Collins T, Maskall J, Lea J, Lunt P, Davies S (2012) Enabling remote access to fieldwork: gaining insight into the pedagogic effectiveness of 'direct' and 'remote' field activities. J Geogr High Educ 36(2):197–222. Retrieved from http://oro.open.ac.uk/30225/

Whalley WB (2014) Alport 'Castles' landslips Quarry, Derbyshire, UK. Retrieved from http://gigapan.com/gigapans/76397/

Wolniewicz P (2014) SedMob: a mobile application for creating sedimentary logs in the field. Comput Geosci 66:211–218

YouTube (2014) Fieldwork video. Retrieved from http://www.youtube.com/watch?v=2baM2JUEI8M

Chapter 9
Conclusions and Recommendations

Abstract In this final chapter we draw together pedagogic and practical aspects that have been developed throughout this book. We bring together the findings from some of our research into the benefits of Technology Enhanced Learning in the field. We also provide additional fieldwork and technology resources, and several series of 'Top Tips' for practitioners wishing to take these topics further and integrate technology into their own fieldwork teaching practice.

Keywords Fieldwork pedagogy · Fieldwork practitioner · Fieldwork resources · Innovative fieldwork · Fieldwork skills

9.1 Conclusions

This book is a collation of good practice accounts of technology enhanced innovative fieldwork teaching. By drawing these pedagogic ideas and technological advances together through a set of examples, practitioners can follow or use them as a starting point for their own innovations in HE fieldwork teaching in a range of allied disciplines. The inherent flexibility of the iPad as a tool to support these various innovations has been shown many times over—but the pedagogic need has always driven the introduction of technology to field teaching. Table 1.2 illustrates the potential linkages between our senses and the functionality of an iPad to demonstrate how the tablet can be employed for a series of standard educational tasks.

As evidenced in this book the 'desktop, laptop, netbook' computing power can now be substantially replaced by 'tablet computing' in the post-iPad era, especially for out-of-classroom' activities. As discussed as the start of the book (Sects. 1.7 and 1.8); mobile technologies such as tablet computers (Whalley et al. 2014) and smartphones (Welsh and France 2012a) are now an important, versatile component of a student's personal learning environment (PLE) (Fig. 1.1). Due to their inherent portability, they provide an extension of computing power outside of traditional (digital) learning spaces. Further, a Group Learning Environment (GLE) can be formed through student collaboration using mobile technologies, either synchronously or asynchronously. While the advent of 'the cloud' for working and developing students' knowledge networks has been fundamental in enabling the use of tablet technology in a wide variety of learning spaces. This can be summarised as shown in Fig. 9.1. We expect to see convergence of some ideas as computing power becomes mobile. For example, VFTs and VFGs may be used in the classroom or lab as preparatory sessions but extended through mobile devices in the field for collecting data with adjunct applications such as mapping tools, GIS and *Google Earth*. The increased processing power, storage capacity, increased resolution and maintained long battery life shows that tablets, at long last, provide truly mobile computing.

One of the main attributes of iPads and iPhones is that they are effectively mobile computers with data input, storage, processing and output devices all rolled into a very mobile package that can be used by anyone, anywhere at any time. For once however, the mystique of 'computer' is not appended and this also contributes to their functionality; people are no longer concerned with the item but rather what

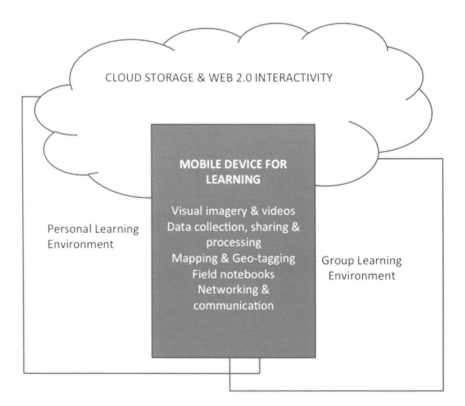

Fig. 9.1 Ways that mobile devices can facilitate the interaction between PLEs and GLEs via cloud storage and Web 2.0 interactivity

9.1 Conclusions

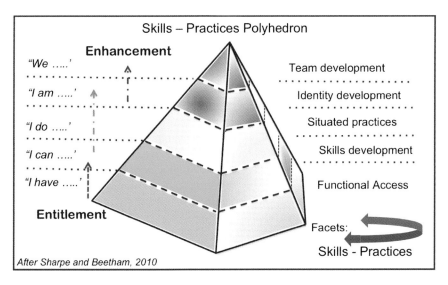

Fig. 9.2 Undergraduate skills-to-practices development schema modified after Sharpe and Beetham (2010) by adding a top tier, Team development' and with mutiple facets-skills

they can do. In this book we have brought some of these aspects together within a pedagogical framework and we would like to summarise this in Fig. 9.2.

Figure 9.2 adapts the general model of digital literacy of Sharpe and Beetham (2010) into a more general framework. It is a pointed polyhedron because it tapers from general entitlement ('I have access to the Cloud') through basic skills to situated practices within, for example, lab work or, especially in our case, fieldwork. We have added a 'We ...' or team element at the apex because the use of mobile devices allows exchange of information between students and practitioners that may well be necessary for problem solving (or employability). The tapering faces of the polyhedron depict skills or practices, such as digital literacy, information literacy, numeracy, spatial awareness, modelling, analysis etc. of whatever discipline is being undertaken. The gradation of colour/tone suggests that not all students at any time will have mastered the necessary skills and practices and that both tutors and students need to be aware of the progress in development.

We suggest that the wide variety of apps now available on mobile devices allow students to gain skills at the lower levels and progress to competencies over time, situations and environments. We consider fieldwork to be of major educational importance in students' learning experiences leading to graduate competencies. However, as we suspected from the outset, if it is possible to use technology in the field to enhance the student experience then it is very likely that the idea or technique, or indeed app, can be used anywhere, from library to classroom and lecture theatre and perhaps more importantly in subsequent graduate-level employment.

9.2 Technology and Employability

Many of the case studies presented in this book illustrate how individuals and groups can personalise their learning environment. It is important for the individual in [higher] education to be able to be familiarise themselves with the opportunity that technology offers in terms of enhanced learning capacity and for that skill set to be carried forward into their subsequent employment. We believe the case studies provided in this book demonstrate a range of techniques and tools in a fieldwork context which will aid familiarisation and increase the users skill-set and confidence.

The application of technology in the case studies is often centred on the use of mobile devices, which clearly can support the development of digital literacy skills, which could aid future employment. Indeed, Whalley et al. (2014) advocates the modern day geographer will have to have an enhanced skill set which captures a range of digital communications and Arrowsmith et al. (2011) reports that graduates will require a robust technological skill sets in the future.

The possibilities of mapping and aligning particular mobile apps to graduate attribute (GA) development in students as been explored by France et al. (in review). They report that students can effectively map mobile apps against predefined GAs and students can perceive linkages between mobile apps and GAs to promote their skills development.

This highlights the fundamental question, is TEL necessary for students to be involved with? We would answer this with affirmation. All of us live in a digital and information rich world. E- and blended learning are more or less out-dated. The more students can gain confidence in using digital devices of all kinds the better placing ownership of learning, especially via problem solving.

9.3 Bring Your Own Device (BYOD)

The BYOD movement started in business but is already gaining pace in education (Johnson 2012) with specific calls for tablet usage (Thiruvathukal 2013). There is not yet ubiquitous ownership of smartphones or tablet devices amongst the student population, but an increasing number of students now own them (Welsh and France 2012a, b). However, Woodcock et al. (2012) reported that many of these students are unaware of the potential of smartphones to support their learning, but importantly, they are interested and open to the possibilities they provide as they become familiar with using them for a range of educational purposes. As such, students will increasingly be prepared to bring device usage into their educational space, not only for fieldwork, but as general replacements for bigger, less convenient laptops with a shorter battery life and so on. It is therefore our responsibility as fieldwork practitioners to help support our students in accessing the learning support that these devices can offer. Institutional support is required in order to establish decent Wi-Fi provision and staff training, but the educational opportunities are vast. Consideration does need to be given to designing activities that remain inclusive for those students without devices and utilising apps that are available across all platforms, but this area of education is only going to expand.

In this final section we bring together some of the main conclusions from our research. We have provided some 'Top Tips' for various aspects of fieldwork and how it can be implemented by, mainly, mobile technologies.

9.4 The Enhancing Fieldwork Learning Project—Conclusions

As a multi-disciplinary team our individual experiences of using technology in fieldwork were developed and brought together through the Enhancing Fieldwork Learning project (EFL 2014). During this 3-year Higher Education Academy-funded project we ran a variety of events to demonstrate how technology, and iPads in particular, can enhance student experiences of fieldwork. In our showcase events we demonstrated technology linked to student experiences in fieldwork, saw demonstrations of other colleagues' work and discussed pedagogic as well as practical issues. We also asked for case studies of the use of technologies in fieldwork and some of them provide the basis for this book. We concentrated on the use of tablets and apps as these are an emerging technology and fulfil the role of 'a computer in the field'; but they also go a long way beyond this in a variety of ways that we had not expected when we started the project.

These summary boxes were originally published in the Final Report of the EFL project (France et al. 2014). These summaries were written for practitioners who wish to understand the potential benefits to themselves and to their students of using technology in their fieldwork practice. We describe the potential barriers they may encounter and the perceptions of students having tried using technology in the field. We also describe a recent study into fieldwork provision in the UK in the Biosciences and GEES disciplines.

---

1. Why introduce technology to fieldwork teaching?
A survey of 76 practitioners highlighted the following main pedagogic reasons why technology was introduced to their field courses:

- To efficiently process data (63.2 %)
- To develop students' technology skills; both in general information & communication technology (ICT) skills and in subject-specific methods and techniques (48.7 %)
- To facilitate post-fieldwork reporting (11.8 %)
- To enhance the fieldwork experience (10.5 %)
- To facilitate communication (10.5 %)

(Welsh et al. 2013)

2. Are there barriers when introducing technology to fieldwork?
A survey of 79 practitioners provided information about the barriers they have faced when introducing technology to their fieldwork teaching:

- Cost—in terms of buying and maintaining the equipment, concerns over insurance and risk of loss or damage to equipment (67 %)
- Reliability and durability of the kit in wet/muddy field conditions (21.5 %)
- Concerns over staff competence/willingness (20.3 %)
- Concerns over student competence or willingness to use high tech equipment in the field (12.7 %)

(Welsh et al. 2013)

3. What sort of tasks were students asked to do using iPads in the field to support their learning activities?
Six fieldtrips across a range of disciplines were conducted using iPads to support learning activities in the field. These fieldtrips were studied as examples of good practice. 173 students were surveyed to find out the main tasks they used the ipads for:

- Taking photographs (other than geotagging) (83 %)
- Mapping/Geotagging (71 %)
- Browsing the web for information (62 %)
- Recording video (47 %)
- Editing videos/photographs (42 %)
- Recording data in Numbers (35 %)
- Recording audio (34 %)
- Writing text on Pages (32 %)
- Reading PDFs/online text books (32 %)

(Welsh et al., in review)

4. What did the students think?
173 students across a range of disciplines and fieldtrips were asked about their perceptions of using iPads to support their learning on fieldwork. The main benefits/drawbacks are listed here along with the number of students that highlighted them:

Benefits	Drawbacks
Quick/easy (55)	Lack of 3G/Wi-Fi (58)
Recording data/audio/photos/video (40)	Heavy/cumbersome (17)
Mapping/spatial (29)	Unfamiliar tool/software (12)
Multi-tool/all-in-one (27)	Glare in sunlight (8)
Sturdy/robust/waterproof (15)	Cold fingers/difficult to use with gloves (6)
Portable (10)	Concerned about damage (5)
Engaging/interesting/fun/interactive (8)	Vulnerable in rain (5)
Easier than paper/paperless (8)	Data lost over internet connection (4)

(Welsh et al., in review)

5. What is the current state of fieldwork provision in HE in the UK?
A change in HE funding in the UK in 2012 saw a trebling of student tuition fees. Therefore, in the summer of that year we surveyed 30 Bioscience and 27 GEES practitioners in 43 HE institutions about their fieldwork provision for undergraduates. We found that the amount of fieldwork offered was stable and likely to remain at similar levels following the tuition fee rise; however, there was a shift towards making all compulsory fieldwork free to students. (Mauchline et al. 2012, 2013; Welsh and France 2012b)

9.5 Top Tips for Tutors

Top Tips for tutors: Starting out using mobile technology in fieldwork

- Do not be afraid to give it a go: Often the most difficult thing is to actually try it
- Perhaps see if there is a colleague using technology that you can shadow
- Think how technology could improve learning or save time, for instance in relation to data collection and collation in the field
- Start small and simple: think of an application which is not going to mess things up too much if it goes wrong
- Try and visualise the use of the technology from the student perspective, what will their experience be?
- Do not assume the students are familiar with the technology … give them chance to practise before using the tech for real
- If using the technology involves group work, think about how each individual will be engaged
- Have you protected the technology against adverse weather and conditions?

- If you need internet connectivity then check it out, make sure the devices are charged
- Help the students identify and recognise the broad range of skills they are acquiring while using technology on fieldwork
- Make sure you think about evaluating your trial use so you can learn from the experience
- Share your innovative ideas and examples of good practice with colleagues as this can often lead to new avenues to pursue
- Don't use technology for technologies sake
- Remember your intervention with technologies should be driven by pedagogy to enhance the student learning experience

Top Tips for tutors: Applying for funding for mobile technologies in fieldwork

- Purchase a class set that can be shared in small groups
- Apply for local funding sources (Alumni funds etc.)
- Consider funding from Teaching & Learning Centres at the institutional level
- Make a case for Faculty or School funds—describe the value e.g. time saved in data collection/sharing in the field
- Quantify the potential impact of the new kit and the potential reach towards enhancing the student learning experience.
- Justify the need and value for money by estimating the number of students that will benefit
- Describe the need for increased accessibility of new technologies to students to enhance their employability
- Consider a multi-disciplinary application for a shared fieldwork resource across Faculties
- Get supportive quotes from colleagues across the Faculty to support the proposal
- Consider encouraging students to use their own devices (Welsh et al. 2011)

9.6 Final Comments

This book has drawn together some great innovations in field teaching and this enthusiasm for making field teaching and learning motivating and engaging has always been our inspiration. Our collective strength as field practitioners is this desire to share our understanding and knowledge of the environment with our students and to instil a genuine sense of fascination that drives them to want to learn more. We have found that the use of mobile technologies can help to engage our students in their field studies, but the real attraction is the passion and enthusiasm of dedicated practitioners creating a dynamic, exciting learning space in the field.

Further Resources

These resources can provide further information if you are interested in incorporating or reviewing your fieldwork practice, especially if you wish to incorporate technology into your teaching.

Resource	Notes
Enhancing Fieldwork Learning: • Website http://www.enhancingfieldwork.org.uk/ • Pinterest https://www.pinterest.com/efieldworkl/ • Twitter @FIELDWORK_NTF • Audit Tool http://www.enhancingfieldwork.org.uk/ Enhancing Fieldwork Learning Audit Tool.pdf	The EFL Project website has an updated set of resources to support this book. The project Pinterest site has an annotated collection of mobile apps and online resources to support fieldwork teaching and the Twitter account has the latest news feed. There is also a link to the audit tool which was designed to allow practitioners to examine and develop current practice by highlighting strengths and weaknesses.
• Maskall and Stokes (2008) Designing effective fieldwork for the Environmental and Natural Sciences (GEES guide) https://www.heacademy.ac.uk/sites/default/files/gees_guides_jmas_designing_effective_fieldwork.pdf • Butler (2008) Teaching geosciences through fieldwork (GEES guide) https://www.heacademy.ac.uk/sites/default/files/gees_guides_rb_teaching_geoscience.pdf • Peacock, J., Park, J.R. and Mauchline, A.L. (2011) Effective learning in the life sciences: Fieldwork. In: Adams, D. (ed.) Effective Learning in the Life Sciences. Wiley-Blackwell, Chichester, pp. 65–90. ISBN 9780470661567	Comprehensive guides outlining how to design effective fieldwork for students in three different disciplines
• Royal Geographical Society and Fieldwork website http://www.rgs.org/OurWork/Schools/Fieldwork+and+local+learning/Fieldwork+and+local+learning.htm • British Ecological Society and Fieldwork website http://www.britishecologicalsociety.org/education/higher-education/enhancing-fieldwork-learning/ • British Geological Survey Higher Education Network website http://www.geolsoc.org.uk/hen	Professional societies information supporting and promoting fieldwork with each respective discipline

(continued)

(continued)

Resource	Notes
• Turner, P (2014) Teaching Geography in a Digital World. Retrieved from *iBooks* https://itun.es/gb/h2TYZ.1	This iBook provides a good introduction into the digital world and teaching
• MoRSE project (Kingston University) related to mobile learning and fieldwork: http://www.slideshare.net/morse1	A series of slideshare presentations around fieldwork and mobile learning
• HEA HUB https://www.heacademy.ac.uk/	Search for 'fieldwork' to find all associated fieldwork resources
• Field Studies Council http://www.field-studies-council.org/publications.aspx	Various resources, publications and guides for fieldwork available here from the Field Studies Council
• Pocket travelling http://www.reading.ac.uk/mmpp/TravelGuideforStudentsandStaffUnbranded.pdf	A guide for students travelling abroad

References

Arrowsmith C, Bagoly-Simó C, Finchum A, Oda K, Pawson E (2011) Student employability and its implications for geography curricula and learning practices. J Geogr High Educ 35(3):365–77

Butler RWH (2008) Teaching geoscience through fieldwork: Plymouth, Higher Education Academy GEES Learning and Teaching Guide.

EFL (2014) Enhancing fieldwork learning project website. Retrieved from http://www.enhancingfieldwork.org

France D, Welsh KE, Powell V, Park JR, Mauchline AL, Whalley WB (2014) Enhancing Fieldwork Learning—NTFS final project report produced for the Higher Education Academy. pp.1-38. Retrieved from https://www.heacademy.ac.uk/sites/default/files/projects/Chester_2010_NTFS_final_report.pdf

Johnson D (2012) On board with BYOD. Educ Leadersh, October pp 84–85

Maskall J, Stokes A (2008) Designing effective fieldwork for the Environmental and Natural Sciences. Higher Education GEES Subject Centre, Learning and Teaching Guide, Plymouth. ISBN 978-1-84102-201-7

Mauchline A, Welsh K, France D, Park JR, Whalley B (2012) Academic fieldwork: six ways to make it work on a budget. Retrieved from http://www.guardian.co.uk/higher-education-network/blog/2012/dec/06/academic-fieldwork-advice-funding-management

Mauchline AL, Peacock J, Park JR (2013) The future of bioscience fieldwork in UK higher education. Biosci Educ 21(1):7–19. Retrieved from http://journals.heacademy.ac.uk/doi/full/10.11120/beej.2013.00014

Sharpe R, Beetham H (2010) Understanding students' uses of technology for learning: towards creative appropriation. In: Sharpe R, Beetham H, de Freitas S (eds) Rethinking learning for the digital age: how learners shape their experiences. Routledge, London, pp 85–99

Thiruvathukal GK (2013) Who needs tablets? We do. Comput Sci Eng 15(1):4–6

Welsh KE, France D (2012a) Spotlight on…smartphones and fieldwork. Geography 97(1):45–51

Welsh K, France D (2012b) The future of higher education fieldwork within GEES disciplines. A report produced for the Higher Education Academy. Retrieved from https://www.heacademy.ac.uk/sites/default/files/the-future-of-higher-education-fieldwork-gees_2012.pdf

Welsh KE, France D, Park JR, Whalley WB (2011) Technology in fieldwork: it doesn't have to cost the earth. Biosci Bull 33:10
Welsh KE, Mauchline AL, Park JR, Whalley WB, France D (2013) Enhancing fieldwork learning with technology: practitioner's perspectives. J Geogr High Educ 37(3):1–17
Whalley WB, France D, Park JR, Mauchline AL, Powell V, Welsh K (2014) iPad use in fieldwork: formal and informal use to enhance pedagogical practice in a bring your own technology world. In: Souleles N, Pillar C (eds) Proceedings of the first international conference on the use of iPads in higher education, Paphos. www.ipadsinhe.org. 20–22 Mar 2014. ISBN 978-9963-697-10-6
Woodcock B, Middleton A, Nortcliffe A (2012) Considering the Smartphone Learner: an investigation into student interest in the use of personal technology to enhance their learning. Stud Engagem Exp J 1(1):1–15

List of Mobile Apps Mentioned

Adobe Acrobat Pro	iBeacon	Python Math
Air Sharing	iCelsius	QRafter
AirDrop	iCloud	Qwiki
AirForShare	iMovie	Readability
Airstash	Inkpad	Readdle
Annotate+	Instagram	Recorder Pro
AppFurnace	iPhoto	sedMob
ArcGIS Online	Irisnote	SeeLevel
Aurasma	iSeismometer	SimpleMind+
Bryce	iSpot	Siri
Bump	iTunes	Situ8
Calibre	iTunesU	SketchBook
Clinometer	Junaio	Skitch
Cloudart	Keynote	SkyDrive
Composer	Lambert	Slideshare
Copy	Layar	Socrative
DataAnalysis	LightMeter	Splice
Decibel	LuxMeter	Splice
DemiBooksComposer	Magnetic Field	Storify
Document5	Memo	SugarSync
DISTO Sketch	Mendeley	Tarragen
Dropbox	Microsoft Office	The Guardian
EasySense	MindMeister	Theodolite
eClicker	MobShow Lite	Total Recall
Educreations	Moleskine	Twitter
Endnote	MovieMaker	Vim
Epicollect	Mscape	Virtual Earth 3D
Evernote	MyScript	Wifi-doc
Everyslide	NearestTube	Wikitude
Excel	Notability	Wordsalad

(continued)

Explain Everything	Notes	Yammer
Facebook	NoteSuite	Zinio
FieldMove	Numbers	Zipcloud
FieldtripGB	Opal	Zotero
Flashlight	Pages	
Fleksy	Panoramio	
Flickr	Paper53	
Flipboard	PDF Expert	
Fotobabble	PDFReader	
GarageBand	Penultimate	
GeoID	PhoneSkope	
GeoSpike	Photobooth	
GIS	PhotoGrid	
GoodReader	Photosynth	
Google Drive	Picassa	
GoogleEarth	Polldaddy	
GoogleMaps	Polleverywhere	
GPS Log	PowerPoint	
iAuthor	Puffin Web browser	

Index

Mobile apps are listed in italics

A
Aanensen, D., 55, 57
Accelerometer, 22, 25, 30, 135
Accessibility, 5, 21, 50, 53, 115
Active learning, 68, 71
Adobe Acrobat Pro, 47, 151
AirDrop, 21, 151
AirForShare, 23, 151
Air Sharing, 29, 151
Airstash, 23, 151
Al-Own, F., 57
Android, 9, 10, 23, 25, 30, 34, 44, 53, 55, 57, 94, 97, 136
Angle measurement, 133, 134
Annotate+, 48, 151
App, 3, 9, 10, 20–25, 28, 29, 33, 34, 39, 47–50, 53, 55, 71, 93, 94, 97, 102, 106, 115, 131, 132, 135, 136, 141, 143
App development, 94, 97
AppFurnace, 44, 151
ArcGIS, 41, 60
ArcGIS Online, 40, 151
Audience response devices, 92
Augmented reality, 41, 44, 106
Aurasma, 106, 151

B
Bartlett, J., 121
Basford, L., 97
Birch, S., 97
Black, A., 97
Blog, 71, 86, 88, 90, 97
Bring Your Own Device/Technology (BYOD/T), 26, 94, 142
Brown, G., 44
Bryce, 41, 151
Bump, 23, 151
Bunting, P., 86

C
Calculator, 25, 135
Calibre, 48, 151
Camera, 19, 22, 30, 34, 39, 41, 52, 53, 57, 65, 68, 71, 75, 83, 103, 106, 121, 127, 129, 130, 134
Catterson, J., 71
Citizen science, 55, 59, 97, 136
Clark, J., 60–62
Clinometer, 25, 131, 133, 134, 136, 151
Cloudart, 93, 151
Collins, T., 25, 115, 116, 121
Communication, 6, 21, 50, 57, 68, 85, 88, 90, 92, 94, 97, 111, 121, 142
Composer, 48, 151
Connectivity, 6, 10, 22, 34, 57, 116
Copy, 23, 75, 86, 88, 116, 151
Culham, A., 97
Curriculum, 41, 94, 116

D
Data
 collection, 8, 24, 25, 34, 50, 52, 53, 55, 57, 86, 94, 97, 103, 116, 134
 processing, 8, 24
 sharing, 23
 storage, 23
DataAnalysis, 25, 151
Davies, S., 25, 121
Decibel, 135, 151
DemiBooks Composer, 48, 151
Dickie, J., 4, 44, 75, 76, 78
Digital camera, 34, 52, 53, 68
Digital notebook, 28, 50, 60
Digital storytelling, 76
Digital surface model, 41
Distance measurement, 131, 133, 134
DISTO Sketch, 133, 151

Document5, 151
Donald, J., 60
Dorn, R., 111
Douglass, J., 111
Drawing, 21, 33, 34, 49, 94, 139
Dropbox, 23, 24, 29, 34, 47, 50, 151

E
Earley, L., 34
EasySense, 25, 151
e-book, 19, 20, 22, 28, 48, 101
eClicker, 92, 151
Educational space, 6, 142
Educreations, 34, 151
e-learning, 4, 30, 75, 111
Endnote, 48, 151
Epicollect, 55, 57, 97, 151
Evernote, 34, 49, 53, 151
Everyslide, 92, 151
Excel, 25, 107, 151
Explain Everything, 34, 152

F
Facebook, 23, 50, 52, 85, 88, 152
Familiarisation, 142
Fearnley, C., 86
Feil, E., 57
Field guide, 18, 28, 102, 109
FieldMove, 134, 136, 152
Field network system, 116
FieldtripGB, 24, 52, 152
Fieldwork, 1, 2, 6, 8, 13, 17, 19, 21, 22, 24–26, 29, 30, 34, 38, 39, 41, 44, 48–50, 53, 59, 65, 68, 71, 76, 78, 83, 85, 86, 93, 94, 102, 103, 107, 109, 111, 115, 116, 121, 127, 132, 133, 136, 139, 141, 143
Fitzgerald, E., 59
Flashlight, 92, 152
Fleksy, 21, 152
Flickr, 52, 53, 152
Flipboard, 86, 152
Fotobabble, 34, 71, 152
Framework for the rational analysis of mobile education (FRAME), 9
France, D., 1, 2, 50, 55, 65, 68, 90, 139, 142, 143
Fuller, I., 1, 68, 132
Funding, 75, 103, 121

G
3G, 22, 29, 116
4G, 22, 116
GarageBand, 30, 80, 152
Geo-referencing, 22, 34, 52, 53, 101
Geo-tagging, 52, 53
Geocaching, 28, 101
Geographical Information Systems (GIS), 34, 41
GeoID, 134, 136, 152
Geoscience, 102
GeoSpike, 50, 152
GigaPan, 102, 113, 129, 130
Global positioning system (GPS), 10, 22
Goodliffe, A., 29
GoodReader, 29, 48, 152
Google Drive, 23, 50, 152
GoogleEarth, 22, 102, 136, 152
GoogleMaps, 22, 152
GPS Log, 34, 50, 152
The Guardian, 48, 151

H
Hackathon, 94
Harrison, C., 55
Hero camera, 127, 130, 131
Huntley, D., 57

I
iAuthor, 48, 152
iBeacon, 25, 151
iCelsius, 25, 135, 151
iCloud, 23, 151
Image, 22, 27, 34, 39, 41, 53, 93, 102, 106, 133
iMovie, 30, 50, 65, 71, 151
Inclusivity, 21
Inkpad, 49, 151
Inquiry learning, 116
Instagram, 23, 151
Internet, 22, 50, 55, 71, 88, 92, 97, 103, 111, 116
iOS, 9, 10, 30, 41, 49, 55, 76, 97
iPad, 8–10, 17, 19, 20, 22, 24, 25, 29, 30, 34, 49, 50, 52, 65, 71, 86, 88, 92, 103, 127, 133, 139
iPhoto, 71, 151
Irisnote, 49, 151
iSeismometer, 25, 151
iSpot, 136, 151
iTunes, 23, 30, 53, 80, 151
iTunesU, 23, 151

J
Jackson, R., 50
Jarvis, C., 4, 44, 75, 76, 78
Junaio, 106, 151

K
Keyboard, 3, 20, 21
Keynote, 28, 39, 47, 49, 151

L
Lambert, 25, 134, 136, 151
Layar, 106, 113, 151
Light measurement, 134
LightMeter, 134, 151
Local wireless network, 24, 115
Lock, J., 94
Lundqvist, K., 97
LuxMeter, 134, 151

M
Maclean, M., 94
Magnetic Field, 135, 136, 151
Magnetic field measurement, 135
Mains, S., 71
Mapping, 22, 53, 55, 86, 90, 113, 140, 142
Mauchline, A, 50, 88, 97
McCloskey, J., 109
McGoff, H., 97
Mediascape, 44
Memo, 49, 151
Mendeley, 48, 151
Microscope, 127, 128
Microsoft Office, 47, 60, 151
Miller, S., 90
MindMeister, 49, 151
Mobile device, 3, 5, 10, 13, 34, 57, 60, 86, 92, 106, 116, 121, 140–142
Mobile learning, 4, 116
MobShow lite, 24, 151
Moleskine, 49, 151
MovieMaker, 76, 151
Mscape, 44, 151
Multimedia, 3, 30, 86, 111
MyScript, 25, 151

N
Nash, R., 55, 94
NearestTube, 106, 151
Nie, M., 76
Notability, 49, 151
Notes, 17, 49, 152
NoteSuite, 49, 152
Numbers, 24, 152
Nuttall, A.M., 109

O
Online learning environment, 86, 88
Opal, 136, 152

Open educational resources, 102
Oppenheimer, P., 97
Outdoor learning, 116, 121

P
Pages, 28, 39, 49, 152
Panoramio, 53, 152
Paper53, 49, 152
Park, J., 65
PDF, 28, 39, 47, 48, 80
PDF Expert, 47, 48, 152
PDFReader, 48, 152
Pedagogy, 2, 3, 9, 116
Penultimate, 49, 152
Personalised learning, 53, 142
Personal learning environment (PLE), 1, 139
PhoneSkope, 128, 152
Photobooth, 71, 152
Photograph, 22, 27, 28, 34, 41, 52, 53, 57, 71, 111, 121, 134, 136
Photography, 27, 41, 71, 121, 127
PhotoGrid, 28, 152
Photosynth, 102, 113, 152
Picassa, 23, 152
PlayStation, 30
Podcast, 34, 53, 65, 75, 76, 78, 80, 82
Polldaddy, 34, 92, 152
Polleverywhere, 92, 152
Porter, P., 103
Poster presentation, 71
PowerPoint, 39, 47, 80, 109, 152
Priestnall, G., 40, 41
Puffin Web browser, 76, 152
Python math, 25, 151

Q
QRafter, 151
QR codes, 92, 93
Qwiki, 86, 151

R
Readability, 48, 151
Readdle, 48, 151
Recorder Pro, 49, 151
Recording app, 97
Remote activity, 121
Remote fieldwork, 116, 121
Research-informed teaching, 103
Resources, 4, 34, 48, 75, 76, 92, 102, 103, 139, 147
Reusable learning objects, 101, 103
Rogers, A., 94
Ruggedised, 18, 50

S

Screen, 19–21, 24, 25, 30, 34, 60, 80, 86, 88, 93, 121, 130, 134
Screencasting, 82
sedMob, 136, 151
SeeLevel, 131, 133, 151
SETT framework, 13
SimpleMind+, 49, 151
Siri, 21, 151
Situ8, 59, 151
SketchBook, 34, 151
Skills, 2, 38, 39, 44, 50, 60, 68, 71, 76, 82, 90, 94, 103, 107, 109, 111, 121, 142
Skills development, 2
Skitch, 23, 34, 49, 80, 151
SkyDrive, 23, 151
Slideshare, 48, 102, 151
Smartpen, 49
Smartphone, 3, 4, 8, 10, 17, 25, 50, 52, 53, 55, 57, 71, 92, 94, 106, 116, 128, 137, 142
Social media, 20, 50, 52, 88, 98
Social networking, 23, 50, 55, 85, 86
Socrative, 92, 151
Sound measurement, 135
Species identification, 94, 97, 136
Splice, 34, 50, 65, 83, 131, 151
Spratt, B., 57
Starting out, 145
Stewart, M., 60
Stimpson, I., 30, 102
Storify, 51, 151
Stott T., 109
Student
 employability, 2, 99, 102, 109, 141
 engagement, 4, 6, 71, 76, 113, 121
 participation, 5, 68, 71, 76, 94
 partnership, 94, 96, 98
Stumpf II, R., 111
Sturdy, C., 94
SugarSync, 23, 151
Surveying, 34, 78, 131, 134

T

Tablet (computer), 3, 6, 8, 20, 21, 27, 55, 76, 88, 121
Tanner, J., 97
Tarragen, 41, 151
Technology-enhanced learning, 6
Telescope, 127, 128
Temperature measurement, 135
Theodolite, 132, 134, 151
Thomas, R., 88
Thompson, D., 25, 107

Thorndycraft, V., 107
Time measurement, 135
Tomlinson, F., 107
Top Tips, 83, 103, 139, 143
Total Recall, 49, 151
Twitter, 50, 52, 85, 86, 92, 151

U

Ubiquitous computing, 2

V

Van Blerk, L., 71
VanCamp, K., 60
Vidcast, 65
Video, 23, 24, 30, 34, 52, 65, 68, 75, 76, 83, 103, 106, 121, 128, 130
Vim, 25, 151
Virtual Earth 3D, 41, 151
Virtual field trip, 101, 106, 111–113, 137
Virtual learning, 111
Virtual learning environment (VLE), 71, 75
Virtual reality, 106
Visual methods, 75
Visualisation, 41, 53, 102
Vodcast, 75

W

Waterproof, 17, 65, 130, 135
Web 2.0, 86, 90, 136
Wells, M., 97
Welsh, K., 1, 8, 53, 139
Whalley, B., 6, 39, 40, 102, 129, 139, 142
White, L., 97
Wi-fi, 10, 22, 27, 49, 116, 121, 131, 137
Wifi-doc, 24, 151
Wikitude, 106, 151
Wills, M., 34, 38
Winterbottom, S., 80, 83
Word clouds, 92, 93
Wordsalad, 93, 151
Wright, J., 116
Wright, P., 55, 116

Y

Yammer, 50, 90, 152

Z

Zilli, D., 94
Zinio, 48, 152
Zipcloud, 23, 152
Zotero, 48, 152